The College Lecture Today

The College Lecture Today

An Interdisciplinary Defense for the Contemporary University

Edited by
Lee Trepanier

LEXINGTON BOOKS
Lanham • Boulder • New York • London

Published by Lexington Books
An imprint of The Rowman & Littlefield Publishing Group, Inc.
4501 Forbes Boulevard, Suite 200, Lanham, Maryland 20706
www.rowman.com

6 Tinworth Street, London SE11 5AL, United Kingdom

British Library Cataloguing in Publication Information Available

Library of Congress Cataloging-in-Publication Data Available
ISBN 978-1-7936-0226-8 (cloth : alk. paper)
ISBN 978-1-7936-0227-5 (electronic)

♾™ The paper used in this publication meets the minimum requirements of American National Standard for Information Sciences—Permanence of Paper for Printed Library Materials, ANSI/NISO Z39.48-1992.

Contents

Preface

This book was a result of an initiative from the Center of Academic Innovation at Saginaw Valley State University. The center's director, Dr. Poonam Kumar, thought it would be beneficial for our faculty to see what were the challenges to those who lecture as a method of instruction and what adaptations were available to overcome them. With support from our Provost, Deborah Huntley, and the Dean of Arts and Behavioral Sciences College, Marc Peretz, the center sponsored a series of workshops throughout the 2017–18 academic year to examine the lecture from the various disciplinary perspectives in the humanities and social sciences.

The main questions we discussed were as follows:

1. What is the lecture? Are there different ways of lecturing? Can the lecture be used to supplement other ways of teaching (and vice versa)?
2. What does the lecture offer to students as a pedagogical technique compared to other forms of teaching (e.g., online, seminars, classroom discussion, service learning)? What are the advantages and disadvantages of the lecture for student learning?
3. Can the lecture be improved as a pedagogical technique; and if so, how? What support can the university provide to faculty members who wish to become better lectures?
4. What is and should be the relationship between the lecture and technology (e.g., online courses, TED, Skype)?
5. How can faculty better communicate to students, administrators, and the public about the value and need of the lecture?
6. What are the unique or specific challenges and values that the lecture plays as a pedagogical tool in your discipline?

After a productive year, it was decided to reach out to other faculty members from a range of universities to share their thoughts and findings about the lecture. The result of our collaboration is this edited volume before you.

I would like to thank not only Poonam Kumar, Marc Peretz, and Deborah Huntley for their support of this project at Saginaw Valley State University but also Patricia Graves and Sponsored Programs at Saginaw Valley State University and Richard Avaramenko and the Center for the Study of Liberal Democracy at the University of Wisconsin for sponsoring me as a visiting professor for the 2018–19 academic year to write and edit this book. I also want to thank all the contributors to this volume as well as Holly Buchanan and the rest of the staff at Lexington Books. Finally, I would like to thank my wife, family, and friends for their support and love for all these past years.

Lee Trepanier

Introduction

The College Lecture Today

Lee Trepanier

Is the lecture still an effective pedagogical technique to teach college students in this age of online education, "flipping the classrooms," and "student-centered education"?

Having been the traditional mode of conveying critical information, history, and theories to students, the lecture is valued by some because it can disseminate information to a large classroom, present material that engages students and elucidates the course material, and provide greater faculty control of the classroom.[1] Donald Blight's classic book, *What's the Use of the Lecture?*, asserts that the lecture is effective at teaching information when compared to other methods of instruction (although less effective in promoting thought about a subject or changing attitudes or behaviors).[2] The popularity of TED talks and Skype presentations also supports those who favor the lecture as an effective method to convey and communicate information to audiences.[3]

However, others have criticized the lecture for a variety of reasons: it promotes passive learning, requires faculty to possess effective speaking skills, and is an antiquated way of conveying information to students raised in the digital age.[4] A recent study published in *The Proceedings of the National Academy of Sciences* compares the lecture with more active forms of instruction and reveal that students taught by active learning scored six percent better on examination than their counterparts in lecture-only classrooms.[5] The invention and adoption of media technologies such as the radio, television, and Internet have also transformed the lecture from one of instructions to performance where entertainment is as important as delivering information to audiences.[6] With the growing pressures of quantifiable assessments for student learning and the growing acceptance of active learning pedagogy, the lecture is increasingly seen as an ineffective and outdated method of instruction.[7]

This book, *The College Lecture Today*, both acknowledges and pushes back against the criticism of the lecture. We believe the lecture still is an effective pedagogical method to instruct students but must be adapted to the new realities of the contemporary university, whether it is the type of student who enrolls at college, the ubiquity of technology in our lives, or recent changes in our understanding of disciplinary knowledge. The need for such a book is especially critical for today, as there has been no new scholarship about the lecture since Blight's book except for those being critical of it.[8]

The College Lecture Today combines both theoretical knowledge and practical suggestions about the lecture in the disciplines of art, English, foreign languages, history, media studies, political science, political philosophy, music, religious studies, and sociology. These chapters are written in an accessible yet scholarly manner that is devoid of academic jargon, so readers can learn about the lecture in a variety of disciplines. The book affirms the value of the lecture while, at the same time, adapting it to the new educational landscape of technology and student-centered learning.

We open our book with a chapter that examines the evidence about the lecture and its relationship to student retention. In "Observed Trends in Lecturing and the Relationship to Student Retention," Westler and French show there is no significant correlation between those professors who lecture and student retention, although there is a significant relationship between those who teach in an autocratic manner and student withdrawals. Westler and French conclude that professors should be more self-conscious of how they teach and consider incorporating methods of active learning in their lectures. The rest of the chapters in this book are in response to Westler and French to show how the lecture still can be a relevant and effective way of teaching students.

The first response comes from the disciplines of art history. In "A Voice in the Dark: The Art History Lecture," Emily Kelley argues that the lecture is still a valuable format for introductory art history courses, particularly when elements of student interaction are part of the delivery. It is through the lecture that the images gain vibrancy where students learn to analyze the objects and apply their knowledge. Kelley concludes with some strategies to engage students during portions of the lecture through reviews, questions, and small group activities.

Our two contributors for English examine how to lecture in composition and theory courses. In "Teaching Writing in Literature Lecture Courses," Taryn Okuma suggests two strategies for improving student writing in lectures. First, the practice of scaffolding writing assignments, and second, the development of a teaching ethos informed by tutoring practice in writing centers. Combined, these strategies can help teach students how to approach writing assignments and build stronger connections to their instructor. With regard to literary theory courses, Elizabeth Rich in her chapter explores in the ways in which faculty have imagined the lecture from its decentered role in mid-century American industrialism to its current phenomenological interpretation. Rich considers techniques for building partnerships between students and faculty that is vital in a world where the value of literature increasingly diminishes.

For foreign language courses, Monika Dix presents insight into the research on teaching culture and language through effective pedagogical practices in the lecture. In her chapter, Dix advocates the 3Ps (products, practices, perspectives), a theoretical approach combining inquiry-based teaching with instructional technology. This technique, along with others, shed light on how to enhance the traditional lecture and create a rich and meaningful environment in which students interact with original material in the target language and develop their own understanding of a foreign culture's products, practices, and perspectives.

The next two chapters look at the lecture in the discipline of history. Hyrum Lewis, "The Continuing Value of Lecture in History Education," argues that, while the lecture is not appropriate for every field, it is highly effective for teaching history. Recent findings in cognitive psychology show that lecture has two major advantages over other methods of teaching history for students: it instills enthusiasm and intrinsic motivation for the subject and it promotes the acquisition of integrated knowledge. But there are conditions that limit when lecture is effective: it must fit the expertise of the instructor and it must be broken up by intermittent exercises that require students to grapple with the material ex-ante (generation) and summarize what they learn ex-post (retrieval).

In "What Lectures Do Well, History Lectures as an Intermediate Pedagogy," Caroline R. Sherman explores the lecture as an intermediate pedagogy, placing it in the middle of the development of the discipline. The young and those without any historical background cannot listen to lectures for long without interrupting with necessary questions—questions that they cannot put aside because the lack of an answer will prevent them from following the narrative. The intermediate learner, however, should have developed the skill of listening to a lecture while also maintaining a running internal commentary on it. Still, the intermediate learner cannot proceed to advanced studies by lectures alone: reading, writing, and discussing are more vital for this, which is why many of the most advanced undergraduate courses and almost all graduate courses are seminars.

From history we move to media studies with contributions from Karen P Burke and Mike Mosher. In "Lecturing and Media Studies in the Twenty-First

Century," Burke explains how integrating project-based curriculum design with the lecture not only improves the lecture but is the key to student and faculty success. Finding new ways to ensure and assess learning by designing curriculum based on timely, real-world problems makes the lecture still relevant in media studies. Mike Mosher in "Campfire, Curator, Deejay: Lectures and 'Lectronic Enhancement" examines how the lecture, combined with electronic media, can be effective in student learning. Specifically, he looks at how the lecture can be used with online platforms such as Blackboard, digital and video imagery, graphic design, and student work projects. The lecture remains central to the process but wrapped in the electronic enhancements to make it more effective.

In political science, John Craig explores the value of the lecture in "The Lecture in Political Science Education: Unpacking a Paradox." As a form of teaching, the lecture has not been without its critics, with both leading political scientists and educationalists highlighting its limitations. Despite this, evidence suggests that it remains the most commonly used form of teaching in the discipline. Craig looks at this paradox by considering the key characteristics of the lecture and identifying its value as a form of teaching in the context of a liberal education. He then examines how the use of the lecture has developed in recent decades, through the adoption of new technologies and its integration with other forms of teaching, and concludes by making the case for value of the lecture as one tool among many.

In "The Lecture in Political Science," I examine the lecture in political science courses and some of the unique challenges each subfield of the discipline poses for it. For each subfield, the lecture needs to be modified to the content being taught. Finally, I end with some practical suggestions of how to lecture in a political philosophy course.

For music, Jamie Fiste argues for the application of gestalt principles to teaching music history and theory. The faculty use of gestalt—the presentation of the larger picture or context of what the student is learning—is critical for students to retain and make sense of what they have learned. Students benefit greatly from faculty who are able to draw connections in inspiring ways in the lecture, reaffirming the value of teacher-centered learning.

The next two chapters look at religious studies. According to Paul Krause in his chapter, because religion touches upon every aspect of culture, the nature of the religious lecture straddles multiple lines: historical criticism in the hope of retrieving information of a past culture, critical scrutiny and engagement with religious stories to recover textual authenticity, and teaching in a value-neutral manner the tenets and concepts found in religious traditions and particular theologies. For Krause, the lecture is ideal for teaching students in the various mediums to which religious studies belong.

"In Creating a Socratic Lecture Space in Religious Studies," Paul Corey examines the gap between a student's spiritual goal and one's professional research interests. The lecture environment can help bridge this divide with a Socratic approach playing a key role. Given the current structure of the university, the lecture hall is still the most viable setting for such dialogue to occur, but it is a setting that must be dramatically reimagined. It must be designed to generate dialogue both *with* and *between* students rather than just professorial monologue.

Our final chapter, "Curious Lectures and Engaged Students: Teaching in a Context of Bureaucracy and Consumerism," is Warren Fincher's analysis of the lecture in the discipline of sociology. To the sociologist, the lecture is a social event, an interaction between a lecturer and a receiving audience for educational purposes shaped by its bureaucratic, professional, cultural, and social structural context. However, the lecturer shares his or her authority in the classroom with an increasing number of external demands: professional accrediting standards, budgeting concerns, and programmatic oversight, while the student has become firmly consumeristic. Thus, as a point of social interaction, the twenty-first-century lecture must attend to three tasks: fulfill bureaucratic dictates, provide a valuable consumer product, and foster disciplinary proficiency in the student.

From these chapters, we see the lecture not only has changed and adapted to new circumstances but the definition of the lecture itself has changed. For some, the lecture is the traditional method of the faculty orally conveying and communicating information to a large group of students; for others, the lecture can be captured in electronic format (e.g., video, online streaming) and enhanced with different types of media (e.g., images, sounds). Some conceive of the lecture as one part among many ways to deliver information to students, while others seek to transform the lecture as a type of Socratic dialogue with students. Regardless of how one defines the lecture, we can see that it is a method of teaching that is far from finished.

Drawing from the humanities and social sciences and from a range of different types of schools, *The College Lecture Today* provides a broad perspective of the lecture in today's university. Although there are some disciplines absent from this book, *The College Lecture Today* provides chapters that are written for nonspecialists and present practical advice of how to lecture in a discipline without sacrificing theoretical knowledge. Neither entrenched in past practices nor discarding the old for the new, *The College Lecture Today* affirms the value of the lecture while, at the same time, seeking to adapt it to the new opportunities and challenges of the twenty-first-century higher education. We hope that this book is not only useful to our readers but also begins a larger conversation about the nature and purpose of teaching college students today.

NOTES

1. J.A. Bowen, "Teaching Naked: Why Removing Technology from your Classroom will Improve Student Learning," *The National Teaching and Learning Forum* 16/1 (2006): 1–5; Mark Lawrence Schrad, "In Defense of the Populist Lecture," *PS: Political Science and Politics* 43/4 (2010): 759–65.

2. Donald Blight, *What's the Use of the Lecture?* (Harmondsworth: Penguin, 1972).

3. Simon Sinek, *How to Deliver a TED Talk: Secrets of the World's Most Inspiring Presentations* (New York: McGraw-Hill, 2013); also see Ronald J. Glogovsky, "A Comparison of Graphic Arts Processes Practiced by Contemporary Industry with Those Taught in Industrial Arts Teacher Education." Master Thesis (Greely, CO: Northern Colorado University, 1970).

4. Charles L. Bane, *The Lecture in College Teaching* (Boston: Gorham Press, 1930); M.A. Albanese, "The Validity of Lecturer Ratings by Students and Trained Observers," *Academic Medicine* 68 (1991): 52–81; Diana Laurillard, *Rethinking University Teaching: A Conversational Framework for the Effective Use of Learning Technologies* (New York: Routledge, 1993); A. O'Donnell and D.F. Dansereau, "Learning from Lecture: Effects of Cooperative Review," *Journal of Experimental Education* 6/2 (1993): 116–25; D. Bruff, "Clickers: A Classroom Innovation," *Thriving in Academe* 25/1 (2007): 5–8; Juan Carlos Huerta, "Getting Active in the Large Lecture," *Journal of Political Science Education* 3/3 (2007): 237–49; W. Berry, "Surviving Lecture: A Pedagogical Alternative," *College Teaching* 56/2 (2008): 102–6; Heather K. Evans and Victoria Cordova, "Lecture Videos in Online Courses: A Follow-Up," *Journal of Political Science Education* 11/4 (2015): 472–82; Daniel J. Mallinson and Zachary D. Baumann, "Lights, Camera, Learn: Understanding the Role of Lecture Capture in Undergraduate Education," *PS: Political Science & Politics* 48/3 (2015): 478–82; James C. Roberts, "Evaluating the Effectiveness of Lecture Capture: Lessons Learned from an Undergraduate Political Science Research Class," *Journal of Political Science Education* 11/1 (2015): 45–60.

5. Scott Freeman, Sarah L. Eddy, Miles McDonough, Michelle K. Smith, Nnadozie Okoroafor, Hannah Jordt, and Mary Pat Wenderoth, "Active Learning Increases Student Performance in Science, Engineering, and Mathematics," *The Proceedings of the National Academy of Sciences*, May 12, 2014. Available at https://www.pnas.org/content/111/23/8410.

6. Norm Friedsen, *The Textbook & the Lecture* (Baltimore: John Hopkins University Press, 2017), 127–38; also see Donelson R. Forsyth, *The Professor's Guide to Teaching: Psychological Principles and Practices* (Washington D.C.: American Psychological Association, 2002).

7. Martin J. Finkelstein, Valerie Martin Conley, and Jack H. Schuster, *The Faculty Factor: Reassessing the American Academy in a Turbulent Era* (Baltimore: John Hopkins University Press, 2016), 7, 12–18, 245. For more about active learning, see the fourth endnote as well as Elizabeth F. Barkley, *Student Engagement Techniques: A Handbook for College Faculty* (San Francisco: Jossey-Bass, 2009); Ken Bain, *What the Best College Teachers Do* (Cambridge: Harvard University Press, 2004); and

Magna Publications, Incorporated, *Active Learning: A Practical Guide for College Faculty* (Madison: Magna Publications, Incorporated, 2017).

8. Books about higher education pedagogy will devote a single chapter on the lecture, if at all, and are only practical in description. For example, Barbara Gross David, *Tools for Teaching* (San Francisco: Jossey-Bass, 2009) and Linda B. Nilson, *Teaching at its Best* (San Francisco: Jossey-Bass, 2010). Friedsen's *The Textbook and the Lecture*, spends most of the analysis on the textbook rather than the lecture and, when on the lecture, writes about it from a purely theoretical and historical perspective rather than a practical one.

WORKS CITED

Albanese, M.A. 1991. "The Validity of Lecturer Ratings by Students and Trained Observers." *Academic Medicine* 68: 52–81.

Bain, Ken. 2004. *What the Best College Teachers Do*. Cambridge: Harvard University Press.

Bane, Charles L. 1930. *The Lecture in College Teaching*. Boston: Gorham Press.

Barkley, Elizabeth F. 2009. *Student Engagement Techniques: A Handbook for College Faculty*. San Francisco: Jossey-Bass.

Berry, W. 2008. "Surviving Lecture: A Pedagogical Alternative." *College Teaching* 56/2: 102–6.

Blight, D. 1972. *What's the Use of the Lecture?* Harmondsworth: Penguin.

Bowen, J.A. 2006. "Teaching Naked: Why Removing Technology from your Classroom Will Improve Student Learning." *The National Teaching and Learning Forum* 16/1: 1–5.

Bruff, D. 2007. "Clickers: A Classroom Innovation." *Thriving in Academe* 25/1: 5–8.

Davis, Barbara Gross. 2009. *Tools for Teaching*. San Francisco: Jossey-Bass.

Evans, Heather K. and Victoria Cordova. 2015. "Lecture Videos in Online Courses: A Follow-Up." *Journal of Political Science Education* 11/4: 472–82.

Finkelstein, Martin J., Valerie Martin Conley, and Jack H. Schuster. 2016. *The Faculty Factor: Reassessing the American Academy in a Turbulent Era*. Baltimore: John Hopkins University Press.

Forsyth, Donelson R. 2002. *The Professor's Guide to Teaching: Psychological Principles and Practices*. Washington D.C.: American Psychological Association.

Freeman, S., Sarah L. Eddy, Miles McDonough, Michelle K. Smith, Nnadozie Okoroafor, Hannah Jordt, and Mary Pat Wenderoth. 2014. "Active Learning Increases Student Performance in Science, Engineering, and Mathematics." *The Proceedings of the National Academy of Sciences*. Available at https://www.pnas.org/content/111/23/8410.

Friedsen, Norm. 2017. *The Textbook & the Lecture*. Baltimore: John Hopkins University Press.

Glogovsky, Ronald J. 1970. "A Comparison of Graphic Arts Processes Practiced by Contemporary Industry with Those Taught in Industrial Arts Teacher Education." Master Thesis. Greely, CO: Northern Colorado University.

Huerta, Juan Carlos. 2007. "Getting Active in the Large Lecture." *Journal of Political Science Education* 3/3: 237–49.

Laurillard, Diana. 1993. *Rethinking University Teaching: A Conversational Framework for the Effective Use of Learning Technologies*. New York: Routledge.

Magna Publications, Incorporated. 2017. *Active Learning: A Practical Guide for College Faculty*. Madison: Magna Publications, Incorporated.

Mallinson, Daniel J. and Zachary D. Baumann. 2015. "Lights, Camera, Learn: Understanding the Role of Lecture Capture in Undergraduate Education." *PS: Political Science & Politics* 48/3: 478–82.

Nilson, Linda B. 2010. *Teaching at its Best*. San Francisco: Jossey-Bass.

O'Donnell, A. and D.F. Dansereau. 1993. "Learning from Lecture: Effects of Cooperative Review." *Journal of Experimental Education* 6/2: 116–25.

Roberts, James C. 2015. "Evaluating the Effectiveness of Lecture Capture: Lessons Learned from an Undergraduate Political Science Research Class." *Journal of Political Science Education* 11/1: 45–60.

Schrad, Mark Lawrence. 2010. "In Defense of the Populist Lecture." *PS: Political Science and Politics* 43/4: 759–65.

Sinek, Simon. 2013. *How to Deliver a TED Talk: Secrets of the World's Most Inspiring Presentations*. New York: McGraw-Hill.

Chapter 1

Observed Trends in Lecturing and the Relationship to Student Retention

Brendon Westler and Eric Michael French

The lecture has increasingly come under scrutiny as an effective pedagogical tool. Images of bored, dispassionate, and passive students, listening (or not) to lectures have led to a rethinking of lectures within the scholarship of teaching and learning (SoTL). Observers have noted of the lecture: "The professor talks (lectures); the students usually listen and occasionally write something in their notes. But something is not right, for not infrequently we observe students having lapses of attention."[1] This rather bleak depiction of the lecture suggests that it is, at minimum, a suboptimal way to deliver content in a classroom.

Critics of the lecture argue that it promotes passive, as opposed to active, learning; that it feels outdated in an electronic and fast-paced age; and that, insofar as education caters to students in order to attract them to our degree programs, the lecture is insufficiently entertaining to fulfill its secondary purpose. Most importantly, the suggestion is that the "'information transfer' model of the traditional lecture does not match what current cognitive research tells us about how humans learn."[2] Lecture is identified as ineffective because it overloads students with information, whereas active learning techniques allow students to actually practice, and therefore better learn, specific knowledge and skills.[3] While lectures are not necessarily inconsistent with active learning principles,[4] they are more likely to reflect instructor-centered rather than student-centered approaches to the classroom. Even when results suggest that lectures are useful for knowledge dissemination, they are found to be less successful than active learning techniques in long-term retention.[5]

But insofar as the above analysis of the lecture as a pedagogical tool hinges in part on the sense that it is boring to the students, curiously the student perspective is all too often overlooked. If students perceive lectures as negatively as the above anecdote suggests, shouldn't we see some sort of indication from the students? Since student evaluations are notoriously unreliable,[6] perhaps there is another metric for gauging student perception of lectures.

This chapter turns attention to a slightly different dimension of lectures: Does lecturing have a correlation with student attrition rates? Putting to the side the question of whether or not lecture is a good match with how people learn or is an effective vehicle for knowledge transmission, we seek here to understand if lecturing has an effect on patterns of enrollment. It would seem that if students who are subjected to lectures in the classroom are bored, disinterested, or otherwise disconnected from their classroom experience, students might actively avoid such courses. Our intention, then, is to bring some new data to bear on the question of the utility of lecturing. We offer here some empirical findings regarding the usage of lecture in the contemporary classroom as well as explore the link between lecture usage and student retention. We report findings from a multi-semester research project wherein we observed courses at various points across the semester. These observations allow us to identify the pedagogical characteristics employed in these classrooms and compare them to rates of student's leaving their courses.

This research reveals several things. First, our classroom observations allow us to provide insight into present pedagogical practices. We find that lectures, despite whatever their negative reputation in the scholarship of teaching and learning, remain a prominent vehicle for delivering lessons. Second, we report that while the use of lectures have no correlation with rates of student retention, the incorporation of active learning techniques are correlated with higher rates of student retention. Lastly, we offer some information regarding the disposition held by instructors in the classroom, providing data which suggests that when the instructor runs a class in an autocratic way there are higher rates of student attrition. In the following sections, we first discuss the use of retention for studying the usage of lectures and the design of our study. Then, we present the findings of our study before offering a brief synopsis and discussion of the results.

WHAT CAN RETENTION TELL US ABOUT LECTURES?

We believe student retention to be a useful metric for evaluating pedagogical choices. Typically, instructors evaluate the effectiveness of their pedagogical choices by determining whether or not the usage of such techniques resulted in their students learning more and better being able to retain that information. This is undoubtedly important in making pedagogical choices—we want to be effective instructors and should therefore use the tools that allow us to be the most effective. At the same time, we believe that considering which pedagogical techniques correspond with higher (or lower) rates of student retention can be a useful piece of information for instructors when designing their courses.

The most common way for evaluating the "effectiveness" of a pedagogical technique is by measuring the extent to which knowledge or skills have been learned or retained over the course of a particular class or course as a whole. The use of a pretest and a posttest, for example, can help show both students and instructors the extent of student learning. Our belief, however, is that this misses out on the possible ways in which pedagogical choices help to determine who and what type of student remain in our classrooms to begin with. That is, if an instructor demonstrates a particular tendency to employ a certain pedagogical technique that is unappealing to a student (for whatever reason), then that student may well simply decide to leave the course. In such a case, the "effectiveness" of the technique is rendered moot by the self-selection of the student. To put it briefly: considering the question of student retention can be a part of the picture of overall effectiveness in that keeping students in the course is the first step in educating them.

This is particularly true when considering the viability of the lecture as a vehicle for education in the twenty-first-century classroom. Amidst the shift toward pedagogical techniques that construct a more "student-centered" classroom lies the underling sense that students who are not actively engaged in the learning process will be less involved in the educative process—and therefore less invested in it. The stereotype of the professor droning on while eyes roll back in heads and no learning is accomplished appears to be precisely the circumstances against which the pedagogical literature has been pushing. And indeed, if students respond to lectures in such a way that they would rather not be subjected to them at all—and therefore withdraw or otherwise leave the course—than this would be useful information to instructors.

STUDYING PEDAGOGY OBSERVATIONALLY

The findings presented here are the result of a long-term research project, conducted over the semesters of spring 2015, fall 2015, spring 2016, and fall 2016. We worked with dozens of undergraduate research assistants to observe pedagogical characteristics of selected courses and then compared them to enrollment statistics for the same courses. This allows us to identify any correlations between pedagogical choices—particularly the usage of lecture in class sessions—and rates of student retention in the classroom. This research was conducted fully aware that there are likely a number of individual-level factors that contribute to whether or not a student decides to remain in a course (e.g., personal health, socioeconomic standing). At the same time, we believe that the information we aim to provide can be useful insofar as instructors may wish to know how their own choices affect student retention. That is, within the areas over which instructors have direct control, what might they be able to do to have a positive or negative effect on retention?

In total, there are forty-three courses in our sample, drawn from across the College of Arts and Sciences at a public research university. The selection of courses was made by filtering the courses offered in a given semester within our target disciplines along two main criteria.[7] First all courses were selected in part because of their low barriers to entry. This does not necessarily mean that they were all "100-level" courses. Rather, the courses were selected because they appeared to admit students without any particular prerequisites. Second, we placed a hard minimum on the number of students in the course. With too few students in the classroom, our observers might disrupt the proceedings. We therefore choose to only observe courses that had at least twenty-two students enrolled (with a median of seventy). Once the courses had been filtered in these ways, we further restricted the sample of courses based on which instructors agreed to be observed and the schedules of our undergraduate research assistants.

As the semesters began, we had our undergraduate research assistants attend and observe a number of class sessions for each course. Each time a class session was observed, there were two research assistants observing and instructors were not notified in advance of the observations. The exact number of observations of the class sessions varied, reflecting the differences in course schedules and activities such as exams or guest speakers (neither of which were observed). All indicators were recorded as a numerical value and then averaged across all class sessions. The result is one score for each course across our tested variables.[8]

When in attendance, our research assistants were asked to complete an observation sheet, giving us information on specific ways in which the class was taught. For this chapter, we report on the following items, all of which were derived from information gathered on the observation sheets: percentage of sessions containing lecture, the average number of active learning components in a class session (by definition, not lecture). We also report here on three indices meant to identify something of the character of the instruction taking place, beyond whether or not there was lecture taking place. Grouping together a number of individual observations, we attempt to show the extent to which the classroom was student centered, the extent to which instructors tried to engage their students, and what we are calling an "autocracy index" to capture the extent to which instructors take on a strongly authoritative or closed-off posture in the classroom. Each of these indices is described in further detail in table 1.1. For each index, all variables are converted to a scale of 0 to 1, allowing for consistency of interpretation across the different

Table 1.1 Description of Variables

Variable	*Description*
Class size	"Class size" refers to the number of students enrolled in the course as of the end of the open drop/add period. This number is used as the basis for the withdrawal rate, D/F rate, and DFW rate.
Withdrawal rate	This is the percentage of students enrolled at the end of the open drop/add period (i.e., "class size") who withdrew from the class after the open drop/add period.
D/F rate	This is the percentage of students enrolled at the end of the open drop/add period (i.e., "class size") who earned a final grade of D or F.
DFW rate	This is the percentage of students enrolled at the end of the open drop/add period (i.e., "class size") who either withdrew from the course after the open drop/add period or earned a final grade of D or F.
Percentage of sessions containing lecture	This variable measures the percentage of observations for a given course in which the observer indicated that lecture was among the activities that occurred.
Average number of active learning components per session	For this variable, five different activities—discussions or debates; simulations or role playing; student presentations; use of student response technology; and group work—qualify as active learning. This variable measures the number of these activities that occurred, on average, per session of a given course.
Student-involvement index	This index combines five different variables, as follows: • Overall level of student involvement in each class session, as rated by the observers; • Frequency with which instructor attempted to gauge student understanding; • Frequency with which instructor addressed students by name; • Frequency with which instructor remained in close physical proximity to students; • Frequency with which students were required to complete work in class. [Note: All of these variables are converted to a scale of 0 to 1 and then averaged, so the overall index value ranges from 0 to 1 and reflects the average value across these five items.]
Engagement index	This index combines three different variables, as follows: • Overall level of instructor enthusiasm during each class session, as rated by the observers; • Overall level of "entertainment value" of each class session, as rated by the observers; • Frequency of student laughter. [Note: All of these variables are converted to a scale of 0 to 1 and then averaged, so the overall index value ranges from 0 to 1 and reflects the average value across these three items.]

(Continued)

Table 1.1 (Continued)

Variable	*Description*
Autocracy index	This index combines seven different variables, as follows: • Frequency with which instructor cold-called students during class; • Frequency with which instructor discouraged sleeping or distraction during class; • Frequency with which instructor discouraged the use of mobile devices during class; • Percentage of class sessions containing lecture; • Frequency with which instructor remained in one fixed location throughout the class session; • Frequency with which instructor attempted to gauge student understanding (reverse coded); • Frequency with which instructor addressed students by name (reverse coded). [Note: All of these variables are converted to a scale of 0 to 1 and then averaged, so the overall index value ranges from 0 to 1 and reflects the average value across these seven items.]

measures. Together, they allow us to comment on how instructors appear to run their classroom and to identify any correlation with student retention.

If the independent variables in this study are the decisions that instructors make in the classroom, whether pedagogical or dispositional, then the dependent variables are the rates of student attrition from a course. To measure student attrition, we draw on enrollment statistics provided to us by the university in order to determine how many students withdrew from our sample classes, as well as how many students received Ds or Fs as final grades. Withdrawals largely capture the phenomenon that we are interested in here. When a student chooses to withdraw from a course after the open add/drop period, they are required to take a W. This means that there is a real cost to such a decision for students and will not be taken lightly. Withdrawing, therefore, should suggest that something has gone wrong. And while we likely cannot say that this is always (or even usually) a result of the pedagogical choices of the instructor, any correlations between pedagogical choice and student withdrawal would be of interest, because other causes of withdrawal should largely cancel out across courses.

The incorporation of D and F grades to our study is meant to capture a slightly different way in which learning has gone wrong and students have been "lost" from the course. Since choosing to take a W requires extra effort on the part of the student, a student who has become lost from the course may just as equally receive a D or an F. Moreover, we assume that a student whose commitment to the course dwindles of the course of the semester is a contributing factor to students receiving such grades. Lastly, we use a combined measure that incorporates the instances of Ds, Fs, and Ws in our courses. While looking at W and D/F rates individually is important, by combining them we are able to get an overarching sense of student attrition.

Our expectations are that when we compare our observational data with enrollment data we will see correlations that reflect the best pedagogical practices as found in the SoTL. In particular, given the skepticism toward the usage of lecture as a viable teaching tool, we expect to see higher rates of student attrition when lecture is used. Similarly, we would expect lower rates of student attrition when active learning techniques are employed. Regarding our three indices, we expect to see lower rates of student attrition when students are more involved in the class and when instructors make an effort to engage with their students. On the other side, we expect to see higher rates of attrition when the class is taught in a more autocratic way.

FINDINGS

Our results suggest several things, both about what instructors actually do in the classroom and what the relationship is between lecture and student retention. First, as seen in table 1.2, we find that, despite being told regularly that one should abandon the practice of lecturing to our students, lecture remains an important component of the contemporary classroom. Our data clearly shows that the courses in our sample regularly employ lecture. Indeed, the

Table 1.2 Summary Statistics[1] (n = 43)

Variable	*Minimum*	*Maximum*	*Mean*	*Median*	*Standard Deviation*
Class size	22	352	96.7	70	77.8
Withdrawal rate	0%	20.5%	7.2%	5.9%	4.9%
D/F rate	0%	27.5%	13.14%	12.7%	7.4%
DFW rate	2.9%	40.6%	19.8%	18.6%	9.8%
Percentage of sessions containing lecture	45.5%	100%	96.5%	100%	9.8%
Average number of active learning components per session	0	1.4	0.3	0.3	0.4
Student-centered index (0–1)	0.07	0.66	0.30	0.28	0.14
Engagement index (0–1)	0.08	0.76	0.45	0.46	0.14
Autocracy index (0–1)	0.21	0.55	0.38	0.36	0.08
Number of observations per course[2]	8	50	18.5	16	8.7

[1]See table 1 for information about these variables and their interpretations.

[2]Two observers attended each class session, so there were two distinct observations per session. The number of class days observed is half of the values shown.

median value of this variable indicates that some degree of lecture occurred in every session of the typical course.

The continued usage of lecture in the classroom can be contrasted to the amount of active learning taking place in the average classroom. In this case, both the mean and median number of active learning activities taking place in the observed classes is 0.3. That is, less than one instance of active learning took place on average during observed classes. Whatever the efficacy of active learning as presented in the SoTL, it remains a relatively infrequent occurrence in classroom.

Turning to the main question of whether or not the usage of lecture is correlated with rates of student attrition, we find that there is no statistically significant relationship either way. This disconfirms our expectations, as we suspected that using lectures would be connected to the loss of students. At the same time, although active learning was comparatively less frequent, we found a correlation ($p < 0.20$) between its usage and W and DFW rates.[9] That is, we do find a correlation between active learning techniques and fewer students being lost from the course. This confirms our expectations (table 1.3).

The three indices meant to reflect disposition in the classroom return interesting results. With the index meant to capture the extent to which the classes were student centered, we expected to see lower rates of student attrition and had the same expectation of classes where the instructor attempted to be particularly engaging. Regarding the former index, we found no statistically significant correlation with rates of student attrition. With the latter, we find a correlation between instructor engagement and lower rates of withdrawal, though only at the 80 percent level. These results largely disconfirm our expectations, as we believed that both indices would result in less attrition.

The strongest results in our study lie with our autocracy index. There we find a correlation with higher rates of student withdrawal that is statistically significant ($p < 0.10$). This confirms our expectation insofar as it seemed that courses that were taught in a particularly autocratic way would likely be linked to higher rates of student attrition.

Table 1.3 Correlation Coefficients ($n = 43$)

Variable	*Withdrawal Rate*	*D/F Rate*	*DFW Rate*
Percentage of sessions containing lecture	−0.0059	0.1683	0.1220
Average number of active learning components per session	−0.2307*	−0.1440	−0.2338*
Student-centered index (0–1)	−0.035	0.1004	0.0529
Engagement index (0–1)	−0.2168*	−0.0324	−0.1396
Autocracy index (0–1)	0.2926**	0.0553	0.1930

*$p < 0.20$; **$p < 0.10$.

DISCUSSION

Our findings suggest that lecture continues to be a primary classroom approach taken by instructors in the contemporary classroom. Whatever the suggestions made by the SoTL and the apparent thrust away from lecture-based models of education, instructors continue to utilize lecture as at least a component of most of their class lessons. Whatever the pedagogical efficacy of lectures as opposed to student-centered or active learning techniques, the fact that instructors continue to use lectures suggests that it behooves us to continue to think through what it means to conduct lectures effectively. Rejecting the discussion of lecture in the SoTL does a disservice to those many instructors who continue to use it. As long as instructors continue to lecture, we ought to determine what characteristics make for the best lectures. Volumes, such as the present one, are therefore both appropriate and necessary.

Then, what does lecture mean for student retention? The findings are somewhat mixed, and we are able to speak only of correlation—meaning we are not able to isolate causal relationships. At the same time, this research offers new and interesting information that can be brought to bear on evaluating lectures as pedagogical tools. We find that the usage of lecture has no statistically significant correlation with our metrics of student retention. This suggests that there might not be a retention-based reason not to lecture—though there may still be other reasons. It also raises further questions about the student perception of lectures. If lectures really are seen as boring by students, with seemingly little knowledge taken away from them, then why do we not see higher rates of student withdrawal? That said, in courses that incorporate active learning, there is a statistically significant correlation with student retention in the W and DFW categories. The suggestion is that active learning might be viewed more positively on the whole and encourages students to remain in the course. While considering questions of student retention may not necessarily dissuade instructors from lecturing, our results suggest that they may wish to consider utilizing some elements of active learning—at least part of the time.

But our other metrics suggest that there may be some stylistic choices that we can make in the classroom, whether lecturing or not, that correlate with higher rates of student withdrawal. In particular, observed classes that demonstrate autocratic characteristics—understood as the instructor attempting to create a disciplined, orderly, and regimented classroom experience—are strongly correlated with student withdrawal rates.[10]

In sum, instructors for whom enrollment in their courses is an important topic can incorporate the information presented here as they determine how they wish to teach their courses. Lectures, whatever their pedagogical efficacy, do not seem to have any effect on the ability of an instructor to retain students in the classroom. Moreover, any individual instructor who chooses to lecture in the face of SoTL recommendations should know that they are not alone as most courses incorporate lecture. At the same time, incorporating elements of active learning may help to increase rates of student retention, meaning that even if instructors wish to primarily lecture, they may be well served by incorporating active learning from time to time. Lastly, whatever the chosen approach to teaching, instructors who take on an autocratic disposition in the classroom appear to have higher rates of student withdrawal. This, it appears, could be more important than the precise pedagogical techniques employed by the instructor when it comes to rates of student withdrawal.

NOTES

1. Joan Middendorf and Alan Kalish, "The 'Change-up' in Lectures," *International Teaching and Learning Forum* 5/2 (1996): 1.
2. Middendorf and Kalish, "The 'Change-up' in Lectures," 2.
3. C. Veronica Smith and LeeAnn Cardaciotto, "Is Active Learning like Eating Broccoli? Student Perceptions of Active Learning in Large Lecture Classes," *Journal of the Scholarship of Teaching and Learning* 11/1 (2011): 53–61.
4. See, for example, Warren Fletcher, "Curious Lectures and Engaged Students: Teaching in a Context of Bureaucracy and Consumerism," in this volume.
5. Adam Wunische, "Lecture Versus Simulation: Testing the Long-Term Effects," *Journal of Political Science Education* (2018). https://doi.org/10.1080/15512169.2018.1492416.
6. Linda B. Nilson, "Time to Raise Questions about Student Ratings," *To Improve the Academy: Resources for Faculty, Instructional, and Organizational Development* 31 (2012): 213–28.
7. Courses were drawn from the following disciplines: political science, history, geography, statistics, philosophy, geology, art, sociology, psychology.
8. For more information on our larger research project and methodology, see Eric Michael French and Brendon Westler, "Keeping Students from Going AWOL: The Link between Pedagogy and Student Retention," *Journal of Political Science Education* (2018). Available at https://www.tandfonline.com/doi/full/10.1080/15512169.2018.1468763.
9. We customarily would not report this level of significance, however, with an *n* of 43 we think this warrants mentioning as a suggestive matter.
10. This could be a spurious correlation, with more difficult courses possibly tending to have both higher withdrawal rates and instructors who teach in a more "autocratic" manner.

WORKS CITED

Michael French, Eric, and Brendon Westler. 2018. "Keeping Students from Going AWOL: The Link between Pedagogy and Student Retention." *Journal of Political Science Education.* Available at https://www.tandfonline.com/doi/full/10.1080/15512169.2018.1468763.

Middendorf, Joan, and Alan Kalish. 1996. "The 'Change-up' in Lectures." *International Teaching and Learning Forum* 5/2: 1–5.

Nilson, Linda B. 2012. "Time to Raise Questions about Student Ratings." *To Improve the Academy: Resources for Faculty, Instructional, and Organizational Development* 31: 213–28.

Smith, C. Veronica, and LeeAnn Cardaciotto. 2011. "Is Active Learning like Eating Broccoli? Student Perceptions of Active Learning in Large Lecture Classes." *Journal of the Scholarship of Teaching and Learning* 11/1: 53–61.

Wunische, Adam. 2018. "Lecture Versus Simulation: Testing the Long-Term Effects." *Journal of Political Science Education.* https://doi.org/10.1080/15512169.2018.1492416.

Chapter 2

A Voice in the Dark

The Art History Lecture

Emily Kelley

While art historians have been encouraged to embrace active learning techniques, and many recent articles on the scholarship of teaching and learning art history have examined those methods, the lecture remains valuable to art history education, particularly at the introductory level.[1] In fact, because of the range of prior knowledge that students bring to the classroom and the typical lack of previous training in art history, the lecture is a crucial tool in these introductory courses. Through careful delivery and the addition of interactive lecturing methods, the well-crafted art history lecture not only builds students' knowledge of the course material but also helps students bridge gaps and correct misconceptions in their prior knowledge, demonstrates for students the methodology used in art historical analysis, and ignites their passion for the discipline.

THE ART HISTORY LECTURE

For those who have not experienced it, the art history lecture is a unique experience. The lecturer, while typically at the front of the room, is not the focus of attention. Rather, the lights are dimmed and the students concentrate on the image or images before them on the screen. The instructor is the voice that makes these images come alive and creates interplay between history and the image. In his chapter in this volume, Hyrum Lewis writes of the importance of enthusiasm in the history lecture and the significance it has in igniting a passion for the discipline among students.[2] This phenomenon is true in the art history classroom as well. As Julia Sienkewicz explains, the stories told in that darkened classroom and in the lecture-based format are what first drew her, and many others, to art history.[3] Beth Harris and Steven Zucker, creators of the popular SmartHistory videos, similarly observe that it was through watching other instructors explain objects in the classroom that they developed the idea for their series of videos.[4] These videos, which discuss works of art in a conversational tone, bring the works to life before a student's eyes and provide them with an intimate experience of the objects.[5] Students who watch the videos, as Harris and Zucker explain, indicate that the videos provide "a deeper understanding of the subject" particularly as compared to the textbook.[6] While the conversational style of their videos is not typically present in an art history lecture delivered by a single individual, the casual tone of the delivery is not unusual.[7] In fact, in many ways the art history instructor, like the narrators in these videos, becomes less of a lecturer, in the traditional sense, and more of a storyteller, weaving together the narrative of the images shown and the story behind them. Harrington and Zakrajsek discuss storytelling as one of the effective and memorable methods of lecture,[8] explaining that, despite the reputation of lecture as a passive teaching method, "a good story can result in a very actively engaged listener."[9] Although the entirety of an art history lecture is not structured as a story, portions of it can provide a unique opportunity to engage students by sharing the stories of the art—either its creation or the narrative that the image shows—with the students. Depending on the image, these can be quite engaging and even funny to the student audience, helping them connect with and learn the material in part by remembering the accompanying narrative.

Despite the engaging methods I have just described, the traditional impression of an art history lecture is one where a professor walks into a large auditorium and dims the lights; then the slide projectors whir to life, projecting images on the screen as the low hum of that machinery accompanies the monotone voice of the instructor. Regardless of discipline, this idea of the monotone, unending monologue is the stereotype of the lecture, casting it in an unnecessarily negative role. What I discuss here as "lecture" is far from that traditional structure but instead embodies a more interactive approach.[10] I am describing lectures that for segments of the class period can be a monologue by the instructor, conveying information so that introductory students can gain the foundation necessary to perform higher-level activities in the course. However, those segments of more traditional lecture are punctuated by dialog between instructor and students, review exercises in small groups, and so on—all types of reflection exercises that can help enhance student learning in a lecture-based course.[11] As active learning has taken on an increasingly prominent role, lecture-based instruction has its place in a classroom where a large amount of basic knowledge needs to be conveyed and organized and where methodology needs to be modeled,[12] all tasks important to the introductory art history course.

ESTABLISHING GROUND WORK IN THE ART HISTORY CLASSROOM

One of the primary reasons that the interactive lecture is appropriate for the introductory course is because it provides critical information about the course material and discipline methodology that students typically have no exposure to prior to their first art history course. Studies show that prior knowledge can either "help or hinder [student] learning."[13] Unlike many other disciplines that students study in college, the typical student enrolled in an art history course will have never studied the subject in high school and will therefore have little direct prior knowledge of the discipline.[14] They often will not know even basic art historical vocabulary (e.g., column, pillar, foreground) nor necessarily know foundational information about certain art historical periods (e.g., Gothic, Renaissance, cubism) though they may be familiar with these terms from other contexts and bring preconceived notions about them as a result. Teaching the course can be further challenged not only by the *level* of prior knowledge but also by the *range* of knowledge of both art and history that students bring to the classroom. These can vary based on the number and quality of history courses students may have taken and on whether they have previous experience in the studio art classroom, which might give them a good knowledge of art vocabulary. One of an instructor's first goals then with each unit is to assess the students' prior knowledge, fill in any gaps, and correct misconceptions.

One technique for establishing prior knowledge is brainstorming.[15] To do so, I like to begin each new historical period with an open-ended question, such as "What comes to mind when you think of the Middle Ages?" For broader historical periods, I know that students will have some prior knowledge, and perhaps some serious misconceptions. While they may have learned about the Middle Ages from high-school history courses, they may also have learned about it from movies or Renaissance fairs. They may think of knights, dragons, and castles; or they might come to the class with the common misconception that it was a period of total culture decline where art was largely absent from society. Through brainstorming as a class, I can both learn what they know and correct any misconceptions that arise. These corrections to inaccurate information are important since beginning the unit with misunderstandings can lead students to resist new material that contradicts their inaccurate beliefs.[16]

Because students' prior knowledge in an introductory art history course is sparse, helping them to connect their own lives to the course material is another strategy to engage them in the lecture and to help them learn the material.[17] For example, when explaining the ways in which particular saints are selected for a Renaissance painting, I'll ask a student to imagine that he or she is designing the object. Since these Renaissance works often demonstrate a family's affiliation with namesakes and saints who offer professional protection, we create a fictional profession for the student and then pick certain saints that his or her family might select based on career, family names, and other interests. Is this lecture? Technically, yes—I do most of the talking. But does this actively engage the student in thinking about how artists and patrons decide what is included in a work of art? Certainly, it does. Moreover, this type of exercise models higher-tier learning activities for students where they are involved in thinking through the creation of a work of art; it also demonstrates an approach that allows them to relate the process to the other works that they learn.[18]

BUILDING FOUNDATIONAL KNOWLEDGE AND PRACTICING HIGHER LEARNING

While the strategies above provide students relevant background information to understand the connections between art and history during each period, they also are the first steps to building foundational knowledge. In her article applying Bloom's Taxonomy to the discipline of art history, Laetitia La Follette placed this knowledge at the bottom tiers of art historical learning: remembering and understanding. As she explains, at this level students learn to identify certain style traits and relate them to a particular era as well as understand how a particular work of art relates to the historical context in which it was made, the artist's or patron's intention for the work, how it was used, how it reflects broader political or social concerns, the meaning of the people and objects represented in image, and so on.[19] For these phases of learning, students need help grouping and organizing the material, which is a critical role of the instructor's lecture explained further below. Once these areas are learned, students can engage in some of the higher-level aspects of learning: comparing and contrasting images; analyzing them alongside relevant historical, literary, or religious texts; relating new images to those they have learned; and so on.[20] These higher levels of learning, involving application, analysis, and evaluation, require students to first acquire some foundational knowledge.[21] In addition, before students perform higher-level analysis on their own, they need a model and low-stakes practice.

An interactive art history lecture can incorporate aspects of higher-level learning while also teaching novice students new information by presenting the material in sections with breaks for student exercises and questions. Studies show that students understand lecture material best when it is broken into shorter segments with room to practice in between each section.[22] For this reason, I try to limit traditional lecture over new material to fifteen-minute lessons, introducing sufficient examples from the period to establish the main traits and concepts. For example, when discussing prehistoric cave paintings, lecture would include the style traits of those paintings and key-related vocabulary, the materials used to create them, and archaeologists' understanding of the probable function of the paintings, as well as details unique to each example. After each section of lecture, I include periods of student "rehearsal"—that is, an opportunity for students to practice applying what they have just been presented.[23] Because most cave paintings share similar traits, students can "rehearse" what they just learned on a new image by responding to my prompts. Depending on the group and the objects covered, these prompts might include questions like:

- Given what you have just learned about the other cave paintings, what materials do you think were used here?
- What vocabulary terms could you apply to this image?
- How does this fit the style of the other paintings we just learned?
- Based on the other images that we saw; how do you think this was used?

This exercise gives students a chance to reinforce what they have just heard by practicing it and also gives the instructor a chance to check for understanding.[24] For students who struggle with note-taking, this exercise provides an opportunity to revisit their notes shortly after taking them and serves to repeat and reinforce the material at a slower speed.[25] This is particularly useful when done in small groups, first allowing students an opportunity to compare notes and then discussing with the class.[26]

While one might be inclined to say that students should acquire the information given in this initial lecture segment from the textbook prior to class, this, in my experience, has not proved to be a realistic approach. Most students find the reading level and density of information in art history textbooks daunting and/or boring, and they simply will not read it for these reasons. They would rather hear the information in a lecture where they can see the images on a large screen and ask questions of the instructor than read page after page of facts in the book. Based on this, I employ the textbook as a tool to introduce critical background information and reinforce key ideas and works of art, rather than the primary mechanism where I expect students to learn the facts or methodology of art history. In recent years, I have started to assign a short selection of pages that emphasize these areas and to give reading quizzes to both reinforce the importance of reading and to assess student comprehension of the assignment. This repeated exposure to the material through both reading and lecture helps students to learn better since what they read becomes the prior knowledge that they bring to lecture, but their own remarks on the textbook and lecture have told me that they prefer and learn more from the class lectures.

LECTURE AS MODEL: ORGANIZING IDEAS AND MAKING COMPARISONS

One reason students favor the lecture is that it helps them to organize and make sense of the extensive information about the images. As Harrington and Zakrajsek have shown, knowledge of relationships between organizations is important to committing that information to long-term memory and to retrieving it.[27] In this volume, Hyrum Lewis discusses the role that a history instructor plays in teaching students to organize the material that they learn.[28] An art history lecture functions in a similar way and helps students to group like objects and see patterns among the works studied. Both, the way I present course content and the questions that I ask during review are designed to help students make sense of the material in this way. In most cases, students are more successful when they group images together as they study them. For example, learning the style traits associated with cave paintings and then applying that knowledge to each image they learn is more efficient than separately learning the same style traits for each image. Moreover, by grouping the material, they learn the information in a way that they can apply in new scenarios. The ultimate goal of an art history survey course is not that students memorize all they can about specific works of art, but rather that they be able to apply what they have learned about those works to new images that they encounter. As I tell my students, I am pleased if by the end of the course they can walk into a major art museum, look around a gallery, and identify (roughly) the period of art that they are seeing before they read any information about it (though I also hope they read about the works). The lecture is critical to helping students make these connections since the instructor can help students to see parallels between information in a way that textbooks do not.

Lecture is also preferable to textbooks for its ability to help students to make comparisons between materials. Typically, students read textbooks in a linear manner, starting at the beginning of an assigned chapter or section and reading to the end. In an art history text, images are treated through the chapter one-by-one, usually with a section of background information at the beginning of each chapter and some nonlinear boxes of text highlighting particular themes interspersed throughout the chapter. Lamentably, without an instructor's encouragement through methods such as the quizzes mentioned above, students often skip these boxes, which tend to draw together the key themes that unify the material. Of course, textbooks make connections between images, usually with a reference to the previous image by figure number (e.g., see figure 12) but only the most careful reader will take the time to look back at the image mentioned and to really consider the differences. Most, if they are reading at all, will skim over that reference. Likewise, repetition if offered in a textbook will be skimmed or disregarded. In the classroom, however, most students recognize repetition as instructor emphasis of important points or review. Similarly, comparisons of images cannot be as easily ignored when the students are faced with them on side-by-side on the classroom screen. When the lecture includes an interactive element to the review of information and/or the discussion of comparative material, the experience becomes even richer for students.

To reinforce the initial lecture over a subject, subsequent class periods offer further opportunity to help students organize their thinking about the content as well as to check for understanding by having students summarize and review earlier concepts. Studies show that having students review notes in groups helps them learn,[29] and, again, this is a mechanism for adding interactive elements to the lecture and giving students an opportunity to practice with the material. Because an introductory art history course is broken into periods and is image based, I like to begin this type of summary and review by placing two images on the screen for the students to use as the focus of their revision. I choose images that are most representative of whatever period the class is reviewing. Then I give students five minutes or so to reflect on what they remember about the period while using their notes, often using a worksheet to focus their efforts on summarizing key points. This type of activity helps students master material by revisiting their notes and also gives them a chance to fill in gaps.[30] During this exercise, I circulate the classroom checking in with the groups and giving further prompts as needed. After this group review, we come together as a class to discuss what they remember. At this stage, I work to help the students organize their observations. Again, taking the example of prehistoric cave painting, I might ask what they remembered about materials and artistic processes used during the period. Then I might ask what they remember about the style of the period and what vocabulary terms they could use to discuss how the images look. Finally, I inquire about the way the objects were used and what they tell us about prehistoric people. By organizing the review in this way, I reflect the type of organization also emphasized in lecture, helping students to see the relationship and connections between the different images they are learning.

Other aspects of introductory art history lectures involve modeling the type of work that art historians do by comparing and evaluating images and then allowing students time to practice these skills. In my introductory courses, I teach students how to apply their knowledge of art history with an exercise that requires two higher-level learning skills.[31] They are asked to compare and contrast images and to evaluate images they have never seen (hereafter "unknown images") in order to assess the time period and function of that work. While these are the types of activities that students ultimately perform on exams, the interactive lecture allows opportunities to demonstrate this process and to allow students to practice it in a low-stakes environment. In both cases, I model the exercise first. For example, when teaching students to evaluate "unknown images," I begin by showing them such an image that is closely related to a period that we have recently studied and guide them through traits that they can visually observe. Taking for example an "unknown image" that is a Cubist work by Picasso, students might remark on its limited color palette, fragmentation of space, or some identifiable aspect of the subject matter. For the first example, I will often draw the conclusions myself to help them build their comfort level with the new format. Then, I will give them other practice questions and ask them to work in groups before we discuss them as a class. When teaching students comparisons, I follow a similar format, but with more lecture guiding them through the first examples.

The unique aspects of art history as a discipline make the lecture an ideal format for introductory courses, particularly in situations where students bring little or varied prior knowledge to the classroom. Far from the boring, monotone lecture that one might envision, a strong art history lecture can make the images come alive and engage students in the history behind them. Through a more interactive format, instructors can keep students engaged, check student understanding, and provide them with practice in some of the higher-level methodologies of the discipline. Even with the lights dimmed, the art history lecture can awaken in students a curiosity about the images they study and sometimes even a deep passion for the discipline.

NOTES

1. M. Gasper-Hulvat provides an excellent overview of these studies in "Active Learning in Art History: A Review of Formal Literature," *Art History Pedagogy and Practice* 2 (2017): 1–32.
2. Please see H. Lewis, "The Continuing Value of Lecture in History Education," in this book.
3. J. Sienkewicz, "Against the 'Coverage' Mentality: Rethinking Learning Outcomes in the Core Curriculum," *Art History Pedagogy and Practice* 1 (2016): 5–6.
4. B. Harris and S. Zucker, "Making the Absent Present: The Imperative of Teaching Art History," *Art History Pedagogy and Practice* 1 (2016): 5–7.
5. The Smart History videos may be found at: https://www.khanacademy.org/humanities/art-history.
6. Harris and Zucker, "Making the Absent Present," 3.
7. R. Nelson, "The Slide Lecture, of the Work of Art 'History' in the Age of Mechanical Reproduction," *Critical Inquiry* 26/3 (2000): 415–19.

8. C. Harrington and T. Zakrajsek, *Dynamic Lecturing: Researched-Based Strategies to Enhance Lecture Effectiveness* (Sterling, VA: Stylus, 2017), 12–14 and 22–24.

9. Ibid., 13.

10. E. F. Barkley and C. Howell Major, *Interactive Lecturing: A Handbook for College Faculty* (San Francisco: Jossey-Bass, 2018), see particularly 15–28; Harrington and Zakrajsek, *Dynamic Lecturing*, 31–33; C. Howell Major, "Engaging Students through Interactive Lecturing," *NEA Higher Education Advocate* 36/5 (2018): 6–9; L. B. Nilson, *Teaching at its Best: A Research-Based Resource for College Instructors* (New York: John Wiley & Sons, Incorporated, 2016), 194–201. Accessed June 7, 2018. ProQuest Ebook Central.

11. Harrington and Zakrajsek, *Dynamic Lecturing*, 31–32. The authors discuss different strategies for these types of interactive moments during lecture on pages 97–109. Barkley and Howell Major also provide many strategies in their volume.

12. Nilson, *Teaching at its Best*, 188–9.

13. S. A. Ambrose, M. W. Bridges, M. DiPietro, M. C. Lovett, and M. K. Norman, *How Learning Works: 7 Researched-Based Principles for Smart Teaching* (San Francisco: Jossey Bass, 2010), 13; Harrington and Zakrajsek, *Dynamic Lecturing*, 39–50.

14. Sienkewicz also notes this, 2.

15. Ambrose et al., *How Learning Works*, 29; Harrington and Zakrajsek, *Dynamic Lecturing*, 46–47.

16. Ambrose et al., *How Learning Works*, 23–24.

17. Ibid., 17.

18. Laetitia La Follette, "Bloom's Taxonomy for Art History. Blending a Skills-Based Approach into the Traditional Introductory Survey," *Art History Pedagogy and Practice* 2 (2017): 2–11.

19. Ibid., 5.

20. Ibid., 5–6.

21. Ibid.

22. B. LaSere Erickson, C. B. Peters, and D. Weltner Strommer, *Teaching First-Year College Students: Revised and Expanded Edition of Teaching College Freshman* (San Francisco: Jossey-Bass, 2006), 91; Howell Major, "Engaging Students through Interactive Lecturing," 8.

23. D. A. Bligh, *What's the Use of Lectures?* (San Francisco: Jossey Bass, 2000), 46.

24. Erickson, Peters, and Strommer, *Teaching First-Year College Students*, 96–97.

25. Nilson, *Teaching at its Best*, 201–5.

26. Barkley and Howell Major, *Interactive Lecturing*, 24 and 274–7; Nilson, *Teaching at its Best*, 195.

27. Ibid., 40–41.

28. Please see H. Lewis, "The Continuing Value of Lecture in History Education," in this book.

29. Bligh, *What's the Use of Lectures?* 125.

30. See note 26 above.

31. La Follette, "Bloom's Taxonomy for Art History," 4–6.

WORKS CITED

Ambrose, S. A., M. W. Bridges, M. DiPietro, M. C. Lovett, and M. K. Norman. 2010. *How Learning Works: Seven Researched-Based Principles for Smart Teaching.* San Francisco: Jossey Bass.

Barkley, E. F., and C. Howell Major. 2018. *Interactive Lecturing: A Handbook for College Faculty.* San Francisco: Jossey-Bass.

Bligh, D. A. 2000. *What's the Use of Lectures?* San Francisco: Jossey Bass.

Erickson, B. L., C. B. Peters, and D. Weltner Strommer. 2006. *Teaching First-Year College Students: Revised and Expanded Edition of Teaching College Freshman.* San Francisco: Jossey-Bass.

Gasper-Hulvat, M. 2017. "Active Learning in Art History: A Review of Formal Literature." *Art History Pedagogy and Practice* 2/1: 1–32. Available at https://academicworks.cuny.edu/ahpp/vol2/iss1/2.

Harrington, C., and T. Zakrajsek. 2017. *Dynamic Lecturing: Researched-Based Strategies to Enhance Lecture Effectiveness.* Sterling, VA: Stylus.

Harris, B., and S. Zucker. 2016. "Making the Absent Present: The Imperative of Teaching Art History." *Art History Pedagogy and Practice* 1/1: 1–7. Available at https://academicworks.cuny.edu/ahpp/vol1/iss1/4.

Howell Major, C. 2018. "Engaging Students through Interactive Lecturing." *NEA Higher Education Advocate* 36/5: 6–9.

La Follette, L. 2017. "Bloom's Taxonomy for Art History. Blending a Skills-Based Approach into the Traditional Introductory Survey." *Art History Pedagogy and Practice* 2/1: 1–11. https://academicworks.cuny.edu/ahpp/vol2/iss1/3.

Lewis, H. 2019. "The Continuing Value of Lecture in History Education." In Title of Book, edited by Lee Trepanier, Lanham, MD: Lexington Books.

Nelson, R. 2000. "The Slide Lecture, of the Work of Art 'History' in the Age of Mechanical Reproduction." *Critical Inquiry* 26/3: 414–34.

Nilson, L. B. 2016. *Teaching at its Best: A Research-Based Resource for College Instructors.* New York: John Wiley & Sons, Incorporated. Accessed June 7, 2018. ProQuest Ebook Central.

Sienkewicz, J. 2016. "Against the 'Coverage' Mentality: Rethinking Learning Outcomes in the Core Curriculum." *Art History Pedagogy and Practice* 1/1: 1–14. Available at https://academicworks.cuny.edu/ahpp/vol1/iss1/5.

Chapter 3

Teaching Writing in Literature Lecture Courses

Taryn Okuma

The introductory survey of literature is a common feature of most English curricula and, at many colleges and universities, these surveys are also taught using the lecture format in order to accommodate higher enrollments of students who may need the course to fulfill a major, minor, or general education requirement. As a result, the "Intro to Lit" course serves as an important gateway course that provides foundational content knowledge for novice literary scholars and nonmajors who are new to the field, while also instructing these same students in the written conventions of the discipline.

Although there are many advantages to the lecture format as an approach for teaching, the lecture is also an unideal format for writing instruction.[1] If good writing pedagogy depends, in part, on more individualized instruction, the lecturer is already at a disadvantage, for she finds herself responsible for instructing a large number of students, which limits her in practical ways when it comes to providing feedback and holding writing conferences, for example.[2] Thus, writing, while crucial, is not the primary pedagogical objective for the instructor or the student. The result is that Intro to Lit lecturers often find themselves focusing more on content knowledge and resigning themselves to the fact that they can only provide a cursory introduction to writing in the discipline.

Yet, even if it isn't possible—or desirable—to transform the lecture into an ideal writing classroom, it is important to consider some ways in which we, as instructors, might optimize the lecture format to make our writing instruction more effective for and beneficial to students. In many Intro to Lit lectures, some class meetings are devoted to lectures (usually given by the designated instructor for the course), while another portion of meetings are structured as discussion sections (usually facilitated by a teaching assistant). This combination of formats allows for content knowledge to be efficiently conveyed through lectures, while also providing regular opportunities for dialogic, active learning in a smaller, more individualized class discussion. Similarly, by adapting a combination of pedagogical strategies from the composition classroom and the writing center, instructors can mitigate the lack of individualized attention that they can give to student writing in lecture courses. While none of these practices are unique or innovative, they are approaches to teaching that lecturers can lose sight of as we strive to meet the content-based goals of our classes. This chapter suggests two strategies for improving student writing in lectures: first, the practice of scaffolding writing assignments, and, second, the integration of informal writing into the course. Combined, these strategies can help teach students how to approach complex assignments and develop long-term goals as writers, two general learning skills that are crucial to student success.

SCAFFOLDING WRITING ASSIGNMENTS

A fairly conventional approach to assigning writing in Intro to Lit lectures involves a midterm essay and a slightly longer end-of-term final essay, both of which require students to engage in literary analysis. However, this basic format can be modified in order to better address the writing needs of the diverse body of students who regularly enroll in lecture courses by adapting and adopting practices from the composition classroom.[3] The approach outlined here is the scaffolding of writing assignments, which is accomplished by breaking up writing assignments into multiple, sequenced assignments.

A common writing objective for many Intro to Lit lectures is to introduce students to the basics of literary analysis. Students usually write essays that engage in close reading analysis and require them to construct an argument informed by a text-based interpretation of form and content. For many students in an Intro to Lit lecture, their primary experiences writing essays for English classes were in high-school AP English classes and in first-year composition classes. In both cases, students learn transferrable writing skills, but they are not necessarily appropriate to college-level writing or discipline specific.[4] The challenge for the Intro to Lit lecturer is to provide this foundational writing training for a large group of students who occupy the full spectrum of writing experience, from experienced writers with varying degrees of exposure to the discipline, to novice writers who have taken few writing courses in any discipline. Simultaneously, the lecturer must account for the variety of reasons why students are enrolled in the lecture, which also affects performance and learning; nonmajors may have more general, localized goals for taking the course, while majors have long-term goals related to mastering the necessary tools for their field of study. The problem is finding productive ways to meet student writing needs while also attending to the limits of what can be practically accomplished within a semester-long introductory course by a lecturer who, even with assistance, is vastly outnumbered by her students.

Sequenced Assignments

The traditional literary analysis essay assignment may seem straightforward, in that most college students are familiar with the idea of a thesis-driven argument and textual analysis. Yet, students still have difficulty with formulating effective theses and engaging in detailed, critical readings of literary texts. Breaking up the traditional essay assignment highlights the different elements that are foundational to strong literary analysis while providing more opportunities for feedback, which makes learning how to write in the genre more manageable for inexperienced writers, while also reinforcing and challenging the skills of experienced writers.[5] Sequenced assignments also have the added practical benefit of discouraging academic dishonesty by "engaging students in the process of writing, which produces materials such as notes, drafts, and revisions that are difficult to plagiarize."[6]

One example of a sequenced writing assignment is the midterm essay that I assign and involves the following components, reproduced here, with due dates, from my most recent syllabus for the course:

- Close Reading Assignment: For this assignment, choose *a single sentence or phrase* from one of the texts discussed in lecture last week and provide a close reading analysis in 250 to 500 words. Be attentive to word choice, punctuation, context, and so on. If you consult a dictionary, use only *The Oxford English Dictionary*. Due 1/19.
- Thesis Statement Assignment: For this assignment, write a working (draft) thesis statement for your midterm essay, as well as a brief explanation (250 words) of how you think you might support your thesis with close reading analysis and some of the potential weaknesses of the thesis. Due 2/9.
- Draft for Peer Review: For this assignment, you will need to bring a working draft of your essay to class for peer review. Your draft can take any form you'd like (outline, list, full draft, etc.), but must include, at a minimum, a working thesis and the quotes that you plan on using to support your claims, as well as at least a sentence for each quote explaining how/why the quote is relevant to your argument. Due 2/16.
- Midterm Essay: Due 3/2.[7]

The sequence of assignments is designed to promote the discipline-specific goal of teaching students to write literary analysis essays, while also working toward a more general goal of coaching students to be better writers. The former goal is achieved through the specific tasks students are asked to carry out at the various stages of the assignment, while the later goal is largely achieved through the way in which the sequence is timed in the course schedule and weighted in the course grades.

Given that students, whether they are majors or nonmajors, often have difficulty with close reading analysis, primarily as a result of inexperience, the first assignment in the sequence is wholly focused on the task. This emphasizes the foundational nature of this skill not only to the essay they will be asked to write, but to the class as a whole. At this very early point in the term (the Close Reading Assignment is due at the end of the second week of classes), students can be alerted to aspects of their close reading practice that need to be refined and immediately experience writing as an integrated part of the course.

The Thesis Statement Assignment is the first assignment in the sequence that is specifically oriented toward the midterm essay in its final form. While the Close Reading Assignment has a more general goal of promoting a skill necessary to the genre of the literary analysis, the second assignment seeks to get students thinking about and actively working on the essay almost a month in advance of the final due date. Again, part of the rationale for this assignment in the sequence is to highlight and provide an opportunity for feedback on a crucial component of the essay at a moment early enough in the writing process and the term to give students time to refine and revise their work.

The Draft for Peer Review is the element of the sequence that is already most commonly incorporated into these types of classes, and also the assignment that is probably perceived as the least helpful by students. Peer review is often much lamented by students precisely because it does not provide them with instructor feedback and students can feel either inadequate to the task of advising their fellow students or cheated by the experience of receiving comments that they do not recognize as authoritative or informed.[8] The effectiveness of peer review as a learning experience can be modified by any number of approaches to how a lecturer organizes the task (e.g., by assigning groups or not), and the technique that connects this assignment to the other tasks in the sequence is presenting the task as a Directed Peer Review (see the appendix for an example of a Directed Peer Review worksheet). With Directed Peer Review, students can benefit from having the qualities and elements of a successful essay explicitly defined for them, regardless of whether or not their peer feedback is helpful. As a result, students still benefit from the experience of reading and interacting with others' writing, yet there is less of a burden placed on students to "figure out" how to evaluate the writing in front of them. After engaging in Directed Peer Review, students have guidelines that they can apply to their own work in order to engage in self-assessment, as well as guidelines and a piece of writing that they can take to office hours or the writing center.

The Importance of Timing the Sequence

This secondary benefit of Directed Peer Review (the student has produced some writing that can be discussed outside of peer review) is reason why the timing of the sequence of assignments is crucial. It may seem excessive to start the process of writing the midterm essay four (or more, if one recognizes the Close Reading Assignment as part of the assignment) weeks in advance, but creating an assignment that unfolds over a longer period of time has multiple benefits for students. First, it can relieve some of the anxiety that attends writing for many students, whether they are experienced or not, by providing feedback at crucial points in the process. Students can be assured of whether or not they are "on the right track" at early and later stages, whether they are formulating their ideas or putting all the pieces together for a draft.[9]

Second, a sequenced assignment that unfolds over time allows students more opportunity to take advantage of other writing resources that are available to them, but often go underutilized due to time restraints. Although students may not visit office hours or take their work to the campus writing center due to an anxiety about interacting with faculty or a perceived stigma attached to obtaining writing assistance, students often self-report that they do not use office hours or the writing center because doing so requires them to have writing to take in, and, in many cases, this writing is not done until they are up against a due date.[10] With a sequenced writing assignment, however, students have a variety of writings related to their essay that they can take at every stage of the writing process. In fact, the sequenced assignment also makes it easier for the lecturer to incorporate the writing center into the course as a student resource, as an outreach visit to the course can be strategically scheduled and individual assignments can also serve as the basis for special collaborative workshops.[11]

For the final essay assignment, the sequence involves fewer tasks, though it incorporates a different assignment that is designed to synthesize the initial tasks for the midterm essay while also emphasizing the connections between all the writing assignments in the course:[12]

- Proposal: For this assignment, write a working (draft) thesis statement for your final essay, as well as a brief discussion of how your approach to writing this essay will build upon your experience of writing the midterm essay. Identify any questions or concerns you have about your plan for the essay. At least 250 words.

Following this assignment, students write a draft for peer review, and the final essay, and the entire sequence unfolds over several weeks. The prompt for the final essay is initially introduced and discussed five to six weeks before the final draft is due (just after they have received their midterm essays back).

By conceiving of each essay assignment as a sequence of smaller assignments that culminate in a final essay, and, further, drawing a connection between the two essays, such that the midterm essay is a preparatory stage in the process of writing the final essay, students can be encouraged to think about writing for the course as a process that unfolds over the entire term, in which each of the assignments builds upon the previous assignments and is oriented toward a long-term goal. This way of thinking about writing encourages students to widen their focus from achieving grades on individual assignments to working on their development as writers.[13] Framing writing as oriented toward the long-term goal of improving students as writers can also be supported by weighting the grades for writing assignments in order to emphasize development over time. In my Intro to Lit lecture, the shorter assignments (Close Reading Assignment, Thesis Statement Assignment, and Proposal) are each worth 5 percent of the final grade (with the graded emphasis on completing the assignment, with comments focused on providing critical feedback), the midterm essay is 10 percent of the final grade, and the final essay is weighted at 20 percent. In this way, students are less likely to be strongly penalized when they are taking the most initial risk; by the time they are turning in work that will make up more of their overall grade, they have already received feedback and had the opportunity to take advantage of other forms of assistance.

INCORPORATING INFORMAL WRITING

The local pedagogical goal of encouraging Intro to Lit students to take a long-term view of their development as writers is to emphasize the process of how one gradually acquires the necessary knowledge and experience for writing strong literary analysis essays and therefore demystify writing in the discipline.[14] More generally, the goal is to provide opportunities for students to see progress in their abilities in order to boost their writing confidence in ways that can translate to contexts outside of the Intro to Lit lecture.[15] One key way to improve student writing is through frequent practice, and while only a limited number of formal (graded) writing assignments can be practically assigned in a term, it is possible to include more frequent informal (ungraded) writing assignments.[16] Two types of informal writing that can be productively incorporated into lecture courses are discussion board posts and cover letters for essays.

Discussion board posts are a frequent feature of many classes across disciplines. The particular value that they can have in an Intro to Lit lecture is to function as part of the scaffolding for writing instruction. First, students must be encouraged to think about their posts as fertile ground for paper topics and arguments, further emphasizing the connection between the generation and expression of ideas for student writers. This requires that lecturers present the posts as a form of prewriting in which students can test out and work through ideas that might later be refined and refocused for an essay. Second, these posts can also provide lecturers and other course instructors with early and ongoing insight into the progress of writers over the course of the semester. Even if the aim is not to individually tailor instruction based on student postings, it is helpful for the lecturer to be aware of trends in a class and maintain a general sense of student abilities.

Finally, even if lecturers or instructors do not respond to each and every post, the opportunity to respond to or acknowledge student writing remains open. Depending on the number of students enrolled in a lecture and the amount of teaching assistance the lecturer has, it might be possible to quickly read through student responses, either before a given lecture or before a discussion section. After doing so, the lecturer might briefly refer to a post in class (e.g., "Liz highlighted a similar point in her post, which you might take a look at after class" or "Mike raised an interesting question—Mike, do you mind repeating it for us now, since it's so relevant to what we're talking about?") or even just quickly acknowledge some of posts as an aside (e.g., "Kate, John, and Briah all wrote about the symbolism of the house in their posts."). More time consuming, though, in many contexts, also manageable, is responding directly to student posts in written form. Occasional responses can be extremely motivational for students, whether a student receives a response because they are high performing and need the extra feedback to acknowledge their abilities and challenge them to push their thinking to the next level, or because they are struggling in the course and need additional support and encouragement to identify what they need to do to improve.[17] Posts can be easily promoted as another way for students to participate in class, allowing students who are more reluctant to speak up to demonstrate that they are actively engaged, with credit assigned simply for having posted and without the expectation that they will receive any explicit feedback on their work.

A second type of informal writing that can be incorporated into lectures is the cover letter for essays. My Intro to Lit cover letters must include "3 things that went well while writing the essay or that the writer thinks are successful in the essay, 3 things that were challenging or that they would like to improve on, and a short 1-2 sentence reflection on what the process of writing the essay was like." The format is open and invites both formally composed and formatted letters, as well as handwritten bullet-pointed lists. Regardless of format, students can benefit from this exercise in self-reflection and assessment, as well as this explicit framing of their essays as part of a longer process of improving their writing skills.[18] Additionally, while there is no need to present cover letters as documents that will receive a direct response, it is also possible to occasionally respond to them, either in written comments attached to the essay or even as part of a casual conversation with a student.

Providing meaningful motivation for writers in lecture courses might seem impractical, if not impossible, due to enrollment sizes, but this is only the case if one takes the view that motivation is tied to quantity of feedback. It is true that by including informal writing assignments and drawing out the process of major assignments through sequencing, there are more occasions for feedback on student work. However, the point is not to respond comprehensively, but to create the possibility of further engagement and the opportunity to respond as needed. If there is a primary critique that any argument for emphasizing writing instruction in the lecture classroom might encounter, it is that such an emphasis demands more attention, and thus more work and, crucially, time from the lecturer—time which, she simply does not have, regardless of her intentions. In fact, research suggests that limited, yet focused feedback can be more effective for students.[19] Therefore, it is crucial to consider the ways in which the organization and framing of the overall course, as well as the individual feedback given to students, can be oriented toward encouraging and supporting students. With this in mind, not only can lecturers adopt sequenced writing assignments and connect students to resources such as the writing center but can also more fully integrate other forms of writing into the course with the aim of cultivating the habit of writing in our students, while also creating opportunities for more personalized interaction.

APPENDIX

Below is an example of a Directed Peer Review handout that I have used in my Intro to Lit lecture course, "ENG 232: History of English Literature, 1795-Present." In this case, students are asked to respond to these questions as out-of-class work, although a handout of this type could also be used to direct an in-class peer review session.

Peer Review Response Sheet

Write your comments on a separate sheet of paper, indicating at the top both your name and the name of the writer to whom you are responding.

Read through the draft once before answering the following questions:

Introduction

1. Is the main thesis clearly articulated in the introduction? Does it make an argument that connects specific details (i.e., word choice, metaphor, rhyme scheme) to a larger theme/issue?
2. Does the introduction make clear the particular argument(s) that the chapter will be making, or does it just summarize some quality about the passage?
   - Having read the entire essay, does the introduction fit the chapter (i.e., does it provide a logical starting point for the chain of reasoning that exists in the body paragraphs)? If so, explain why, if not, explain what is unclear.

Body Paragraphs

3. For each body paragraph, can the main point be easily identified?
   - Is it clearly articulated in the topic sentence of the paragraph?
4. In each body paragraph, is textual support offered for each argumentative claim?
5. In each body paragraph, how is the evidence linked to the main point of the paragraph?
   - Is each quote followed by a detailed explanation of why the quote is significant or relevant to the argument?
6. Are specific word choices or images (etc.) identified and analyzed?
   - Is it explicit how the paragraph's argument connects back to the overall argument?

Conclusion

7. How does the writer revisit (but not repeat) the major claim of the chapter in the conclusion or offer a new (though not tangential) perspective?

Global Issues

8. In the course of the entire chapter (intro, body, and conclusion), is there any extraneous information (excessive quotation or restatement of the text without analysis, unsupported claims or opinions, etc.)?
9. Does the writer give equal treatment and attention to issues of form and content—does he or she spend the same amount of time analyzing both what the text is saying and how it is saying it?
10. Does the flow of the chapter break down at any point (i.e., does it become hard to follow the chain of reasoning or lose its coherence)?
    - Where? How might you fix it?
11. Does the writer ultimately (in the conclusion) place the thesis/major claim into a broader perspective or context?
12. Does the writer ultimately link the specific details of the passage with the broader perspective or context of the overall text?

13. What has the writer done well in this essay so far? Identify two of the chapter's strengths.
14. Are there any questions at the end of the essay that you would have liked the writer to answer? Preferably, they should be about content or clarification of ideas, but you may also ask about sentence style, word choice, or the sequence of ideas in the essay.

If the Draft is Minimal, Please Respond to the Following Questions:

Thesis:

1. Does the thesis make sense? Why or why not?
2. Does the thesis present an argument that you might find debatable (i.e., it seems plausible, but you'd want to see what textual evidence they found and how they analyzed it before you agreed completely with them), or is it something that seems more or less self-evident upon first reading of the text(s)? Explain your response.
3. Does the thesis go beyond an obvious argument of issues (i.e., "Wordsworth critiques society for being disconnected from nature," "This passage in *Frankenstein* demonstrates the importance of empathy," or "Burke argues that the French Revolution is a threat to tradition")? Does the chapter attempt to address questions of "how/why"? In other words, "*how* and *why* does Wordsworth critique society for being disconnected from society?" and so on. In what sense does the chapter succeed in/fall short of doing this?

Close reading examples:

Please respond to these questions for any quotes that have been identified in the outline.

1. Has the writer chosen a quote that was not discussed extensively in lecture or section? If it was discussed, is it clear that the writer has offered insights and analysis that go beyond what was mentioned in class? Please briefly explain how the writer has or has not done this.
2. Has the writer identified specific words or images of interest in the quote and noted why they are interesting? Could they develop their analysis/interpretation further?
3. Has the writer discussed issues of form (such as stage directions, rhyme scheme, punctuation, line breaks) and how they relate to the meaning of the quote? If not, identify some details they might take note of, though do not fill in the analysis.

4. Are there words or phrases that you noticed in the chosen quote that you think the writer should discuss? Please indicate what they are, though don't offer your own analysis.

NOTES

1. Studies of lecturing pedagogy are careful to qualify their judgments by being attentive to the specific goals of courses when they evaluate the effectiveness of lecturing. C. Mulryan-Kyne, "Teaching Large Classes at College and University Level: Challenges and Opportunities," *Teaching in Higher Education* 15/2 (2010): 175–85; K. L. Dean, and S. Wright, "Embedding Engaged Learning in High Enrollment Lecture-Based Classes," *Higher Education* 74/4 (2017): 651–68.

2. National Council of Teachers of English, "Professional Knowledge for the Teaching of Writing," February 28, 2016. Available at http://www2.ncte.org/statement/teaching-writing/.

3. Though not specific to the lecture classroom, see John Bean's *Engaging Ideas*, particularly Chapter 13. J. C. Bean, "Coaching the Writing Process and Handling the Paper Load," in *Engaging Ideas: The Professor's Guide to Integrating Writing, Critical Thinking, and Active Learning in the Classroom* (San Francisco: Jossey-Bass, Inc, 1996).

4. D. Berrett, "Students Come to College Thinking They've Mastered Writing," *The Chronicle of Higher Education*, March 21, 2014.

5. R. Bruning, and C. Horn, "Developing Motivation to Write," *Educational Psychologist* 35/1 (2000): 25–37.

6. Council of Writing Program Administrators, "Defining and Avoiding Plagiarism: The WPA Statement on Best Practices," January 2003. Available at http://wpacouncil.org/positions/WPAplagiarism.pdf.

7. I have not included the text of my midterm essay assignment, which is a fairly conventional a 3–4-page thesis-driven close reading analysis, because Intro to Lit essay prompts can vary based on instructor and departmental goals and ultimately, my focus here is on the sequencing of the overall assignment.

8. D. Gooblar, "Why Students Hate Peer Review," *Pedagogy Unbound*, March 1, 2017. Available at https://chroniclevitae.com/news/1718-why-students-hate-peer-review; J. H. Kaufman, and C. D. Schunn, "Students' Perceptions about Peer Assessment for Writing: Their Origin and Impact on Revision Work," *Instructional Science* 39/3 (2011): 387–406.

9. It is important to note that the detail and quantity of feedback on student writing is not discussed or prescribed in these suggestions. Enrollments in lecture courses vary widely, from forty to fifty students, up to the hundreds. Discussion of how to manage feedback in a lecture course deserves its own study and consideration, and can be approached in a variety of ways, depending on enrollment, including rubrics, checklists, grading contracts, and so on. The pedagogical strategies discussed in this chapter can be applied in conjunction with any of these feedback approaches and by both lecturers and teaching assistants.

10. T. Okuma, "Understanding Student Perceptions of the Writing Center: A Conversation between a Student, a Writing Center Instructor, and a Director/Professor," *Another Word*, April 29, 2013. Available at https://writing.wisc.edu/blog/understanding-student-perceptions-of-the-writing-center%E2%80%93a-conversation-between-a-student-a-writing-center-instructor-and-a-directorprofessor/.

11. L. Wilson, and O. LaBouff, "Going Beyond Remedial: The Writing Center and the Literature Class," *The Writing Center Journal* 6/2 (1986): 19–27.

12. This modification of the assignments in the sequence is partially designed to introduce a degree of "fading the support provided to the learner so that the learner is now in control and taking responsibility for learning." S. Puntambekar, and R. Hübscher, "Tools for Scaffolding Students in a Complex Learning Environment: What Have We Gained and What Have We Missed?" *Educational Psychology* 40 (2005): 1–12.

13. Bruning and Horn, "Developing Motivation to Write."; R. Ness, "Cultivating a Growth Mindset: Changing Students' Beliefs About Learning," *Another Word*, December 4, 2017. Available at https://writing.wisc.edu/blog/cultivating-a-growth-mindset-changing-student-beliefs-about-learning/.

14. J. P. Lampi, and T. Reynolds, "Connecting Practice & Research: From Tacit to Explicit Disciplinary Writing Instruction," *Journal of Developmental Education* 41/2 (2018): 26–28.

15. National Council of Teachers of English, "Professional Knowledge for the Teaching of Writing"; Bruning and Horn, "Developing Motivation to Write"; J. Mackiewicz, and I. Thompson, "Instruction, Cognitive Scaffolding, and Motivational Scaffolding in Writing Center Tutoring," *Composition Studies* 42/1 (2014): 54–78.

16. National Council of Teachers of English, "Professional Knowledge for the Teaching of Writing."

17. J. Burkland, and N. Grimm, "Motivating through Responding," *Journal of Teaching Writing* 5/2 (1986): 237–47; R. J. Connors, and A. A. Lunsford, "Teachers' Rhetorical Comments on Student Papers," *College Composition and Communication* 44/2 (1993): 200–23.

18. H. Andrade, and A. Valtcheva. "Promoting Learning and Achievement through Self-Assessment," *Educational Theory and Practice Faculty Scholarship* 12 (2009). Available at https://scholarsarchive.library.albany.edu/etap_fac_scholar/12.

19. M. A. Shapiro, "How to Talk with a Student Who Isn't There," *Another Word*, April 21, 2014. Available at https://writing.wisc.edu/blog/how-to-talk-to-a-student-who-isn't-there/.

WORKS CITED

Andrade, H., and A. Valtcheva. 2009. "Promoting Learning and Achievement through Self-Assessment." *Educational Theory and Practice Faculty Scholarship.* 12. Available at https://scholarsarchive.library.albany.edu/etap_fac_scholar/12.

Bean, J.C. 1996. *Engaging Ideas: The Professor's Guide to Integrating Writing, Critical Thinking, and Active Learning in the Classroom.* San Francisco: Jossey-Bass, Inc.

Berrett, D. 2014. "Students Come to College Thinking They've Mastered Writing." *The Chronicle of Higher Education* (March 21). Available at https://www.chronicle.com/article/Students-Come-to-College/145473.

Burkland, J., and N. Grimm. 1986. "Motivating through Responding." *Journal of Teaching Writing* 5/2 (Fall): 237–247.

Connors, R.J., and A.A. Lunsford. 1993. "Teachers' Rhetorical Comments on Student Papers." *College Composition and Communication* 44/2 (May): 200–223.

Council of Writing Program Administrators. 2003. "Defining and Avoiding Plagiarism: The WPA Statement on Best Practices." (January). Available at http://wpacouncil.org/positions/WPAplagiarism.pdf.

Dean, K.L., and S. Wright. 2017. "Embedding Engaged Learning in High Enrollment Lecture-Based Classes." *Higher Education* 74/4 (October): 651–668.

Gooblar, D. 2017. "Why Students Hate Peer Review." *Pedagogy Unbound* (March 1). Available at https://chroniclevitae.com/news/1718-why-students-hate-peer-review.

Kaufman, J.H., and C.D. Schunn. 2011. "Students' Perceptions about Peer Assessment for Writing: Their Origin and Impact on Revision Work." *Instructional Science* 39/3 (May): 387–406.

Lampi, J.P., and T. Reynolds. 2018. "Connecting Practice & Research: From Tacit to Explicit Disciplinary Writing Instruction." *Journal of Developmental Education* 41/2 (Winter): 26–28.

Mackiewicz, J., and I. Thompson. 2014. "Instruction, Cognitive Scaffolding, and Motivational Scaffolding in Writing Center Tutoring." *Composition Studies* 42/1: 54–78.

Mulryan-Kyne, C. 2010. "Teaching Large Classes at College and University Level: Challenges and Opportunities." *Teaching in Higher Education* 15/2: 175–185.

National Council of Teachers of English. 2016. "Professional Knowledge for the Teaching of Writing." (February 28). Available at http://www2.ncte.org/statement/teaching-writing/.

Ness, R. 2017. "Cultivating a Growth Mindset: Changing Students' Beliefs About Learning." *Another Word* (December 4). Available at https://writing.wisc.edu/blog/cultivating-a-growth-mindset-changing-student-beliefs-about-learning/.

Okuma, T. 2013. "Understanding Student Perceptions of the Writing Center: A Conversation between a Student, a Writing Center Instructor, and a Director/Professor." *Another Word* (April 29). Available at https://writing.wisc.edu/blog/understanding-student-perceptions-of-the-writing-center%E2%80%93a-conversation-between-a-student-a-writing-center-instructor-and-a-directorprofessor/.

Puntambekar, S., and R. Hübscher. 2005. "Tools for Scaffolding Students in a Complex Learning Environment: What Have We Gained and What Have We Missed?" *Educational Psychology* 40: 1–12.

Shapiro, M.A. 2014. "How to Talk with a Student Who Isn't There." *Another Word* (April 21). Available at https://writing.wisc.edu/blog/how-to-talk-to-a-student-who-isn't-there/.

Wilson, L., and O. LaBouff. 1986. "Going Beyond Remedial: The Writing Center and the Literature Class." *The Writing Center Journal* 6/2 (Spring/Summer): 19–27.

Chapter 4

Decentering the Lecture as Responding to Material and Historical Changes

Elizabeth Rich

This book addresses the place of the lecture in a changing educational landscape. While it is tempting to think that technologies called "disruptive" to various economic sectors are the reason for change, it is really larger historical forces that affect these changes. In *The Postmodern Condition*, Lyotard says, "To the extent that learning is translatable into computer language and the traditional teacher is replaceable by memory banks, didactics can be entrusted to machines linking traditional memory banks (libraries, etc.) and computer data banks to intelligent terminals placed at the students' disposal" (50).[1]

However, sophisticated databases easily accessed by students and teachers did not precipitate the question of the role of lecturing in the literature classroom. Mid-century pedagogy in literary studies questioned its role and promoted a balance between class discussion and providing lecture notes. In 1953 *College English* asked, "When should the teacher of literature lecture? When allow or provoke discussion?" and dedicated a 1955 issue to the discussion that followed.[2] Inside of this question, one sees a democratization of the classroom. There is a consideration of accepted knowledge but one that expands to include how the knowledge is received by students. Contributors to the 1955 edition of *College English* addressed pedagogies that decentered the lecture from its primary position in the nineteenth-century classroom while foregrounding the conditions of the mid-twentieth-century classroom—the rise of the large lecture classes versus small seminars as a response to a workforce that needed training after high school for an increasingly specialized society.

How the producers and conveyers of knowledge (researchers and professors, respectively) are imagined is at odds with their actual work, and the lecture that is an ahistorical hybrid of the nineteenth century and mid-twentieth century is a persuasive fantasy in popular culture. The lecture lives in various current interpretations in literary texts and film adaptations that fixate on its older out-of-time role. The critique found in late twentieth- and early twenty-first-century popular culture is one that takes issue with a ghost that haunts the imagination. As the lecture changes in the current classroom, actual students and professors are working through increased political interference that determines course content, economic pressures as public universities not only serve private interests by graduating employees but also a transition in which the public university is being parceled out to various "vendors" and turning into a collection of private, for-profit ventures, as well as increased reliance on part-time and temporary faculty. In the current literature classroom, international students are increasingly concerned about their academic status, professors face not just political pressure on the shape of content for their courses but also the erosion of institutional confidence in literary study, and domestic students flee the literature major.

In response to this dynamic, a Modern Language Association panel in January 2018 addressed "Insecurity in the Classroom: Programs, Pedagogy, and Peripateticism" and invited papers to share insights and techniques regarding how to address the anxiety that national immigration debates, looming just outside the classroom door, bring into the classroom itself. Understanding how the public perceives academic work in literary studies is vital to addressing misconceptions that drive students away from the humanities at large and literary studies in particular. It is useful to consider techniques for knowledge building as a partnership between students and professors that is vital for a demanding and increasingly violent world, but these are always done in the context of larger forces that determine enrollment, job security, and the role that literacy plays in a data-driven world.

Historically, the lecture played a central role in the nineteen-century classroom, and in this context it was shaped by the professionalization of literary analysis. The nineteenth century was a time of a solidification of the disciplines that grew and developed through the twenty-first century. Pioneers in biology (Charles Darwin), psychology (Sigmund Freud), literature (Matthew Arnold), economics (Karl Marx and Adam Smith), among other founders of meta-discourses developed the research questions and early methods of today's disciplines. In literary studies, the school of criticism dominated by New Critics included "the central concern of criticism [which] has appeared to mean cutting it loose from its author and from his life as a man with his own particular hopes, fears, interests, conflicts, etc." (Wimsatt and Beardsley 30), suggesting that prior to the early twentieth century, criticism centered on the intention of the author.[3]

A central part of the discipline, which New Critics largely shaped for decades, is to read literature with an eye to the text itself, identifying the process of musing on the author's intention as "a fallacy." A disciplinary orientation will set up assumptions, a common language, and ideas that the neophyte must learn in order to discourse with established practitioners. Disciplinarity, according to Linda Adler-Kassner, finds itself being a point of debate between academics and agents of educational disruption, such as The Lumina Foundation and other groups, who are driving public policy and transforming the expectations at accrediting agencies, such as the Higher Learning Commission (437, 445).[4] Adler-Kassner's well-documented assessment of three fundamental aspects of contemporary education that must respond to the needs of the new posthuman economy (liberal learning, professional education, and disciplinarity) suggests that the curriculum, which is seen as being balanced in the eyes of academics, is being criticized as failing by "educational reformer." This disparate set of views has an impact not only in the curriculum but how instruction is delivered. While it may seem that academics are holding onto antiquated ideas of "transmission of knowledge," one that has thoroughly been redressed in the popular imagination, their practices are not as out-of-date as it may seem.

The media stereotype of the middle-aged, white man teaching canonical literature to a group of restless college students, both having fewer intellectual and more carnal encounters, fails to expose shortcomings of using the lecture in pedagogy in literature classrooms, though it stands firm in the popular imagination. Rather, Timothy L. Carens's "Serpents in the Garden: English Professors in Popular Film and Television" (2010)[5] argues that the sexually predatory white, male professor of canonical literature serves a role in which the skill of interpretation is transformed into sexual tension, a reversal of Freud's notion of sublimation in which sexual tension is translated into professional, intellectual, artistic, or athletic achievement. It is worth addressing this salacious fiction because the popular audience consumes these contemporary archetypes regularly over the course of decades.[6]

Yet, the anxiety that is at issue here is not lust ramped up on power. It is the authority of the transmission of literary knowledge. The professor, having read far more than the students, teaches literature, which Carens differentiates from teaching other humanities courses, such as history where professors lack the impulse to engage in romantic relationships with their students. Rather than impart facts or quantifiable information about literature to students, this professor holds the knowledge at arm's length, judging the students' ability to get close to a sophisticated interpretation, what New Critics articulated in various ways but that suggest that there may be "an ideal reader" who can glean "the 'true' reading of the poem" even while admitting that such a thing cannot exist (Brooks 17).[7] Carens keenly identifies:

> They are drawn to an intellectual activity that encourages them to tap into and display their creative ingenuity. Yet they are put off by the presence of an

> authoritative figure who persuades them to accept ideas that strain their credulity or, worse still, who questions the merit of their own responses. The process of interpreting literature thus generates feelings that find themselves at home in narratives of seduction and betrayal. (12)

Carens goes on to liken the fictitious professor to a priest and says that this "holy" sphere of literature to which the professor can grant or deny access takes on mythic proportions in the symbolic gesture in which the narrative denies sexual attraction for the professor and makes the real object of desire the "forbidden fruit" of the course content itself (15). Even inside the popular imagination that seems to misdirect or misunderstand the college classroom, one can see that the role of the lecture, if embodied in the authoritative persona of the professor, as a means to transmit knowledge has grasped the fact that this trope is a fiction after all.

The posthuman era in which the humanities and their keepers have lost what was assumed to be a clear place with certain authority is really not a significant, recent, or technologically driven change. The search for knowledge is continuous, and the methods of literary scholarship as it is constructed in popular media benefits from early twentieth-century positionalities by poet critics such as Matthew Arnold and T. S. Eliot who saw in literature a secular answer to the nineteenth century's crisis of faith. The speaker in Arnold's "Dover Beach," observing the cold impersonal waves on the shore at night, says, "The Sea of Faith/Was once, too, at the full, and round earth's shore/Lay like the folds of a bright girdle furled./But now I only hear/Its melancholy, long, withdrawing roar,/Retreating, to the breath/Of the night-wind, down the vast edges drear/And naked shingles of the world."[8]

While the nineteenth century witnessed a progressively secular world, redemption for humanity was thought to be found in poetry that worked in the grain of the Western canon. The ostensible power of the poet was to articulate the new modern forces so that poetry and the critics who interpreted it in scholarship and classrooms could translate it to readers as a continuation of rather than break from "Tradition." This need to interpret allusion-laden fractured and complicated forms beckons the same mystique of popular film and fiction. Yet, this view of literature as something that holds onto a sacred set of texts, even using the Catholic term "canon" for select worthy literary texts, changed in the mid-twentieth century to become inclusive of a diverse number of authors and to challenge new critical readings of previously accepted texts. That is, the methods of the discipline changed, but the character of the depiction in the mind of contemporary culture remains in the early twentieth century.

The stagnated image of disciplinary work in literary studies brings with it the figure of a literature professor, lecturing (professing), lodged in an institution. However, academic work has become increasingly precarious. Kieryn McKay and Kylie Brass make the case that the tool that replaces the lecture, the podcast, arrives at a time when universities are cutting their full-time, tenure-track faculty in favor of more flexible instructor positions: "There is a parallel between the appropriation of podcasting technology into the university and the current system of casual academic employment in Australia, in that both the podcase and the casual academic represent 'new' interfaces of outsourced academic labor" (141).[9] They locate the same phenomenon in U.S. colleges and universities, suggesting that this is a structural shift and not one tied to unique national policies: Buxton et al. speculate that "by 2011 American law students will be, in part, basing their choice of law school on the availability of podcast lectures [making the] adoption of podcast lectures [. . .] an act of competitive necessity" (142).

Due to various factors, such as online multitasking behaviors (texting, watching videos, listening to music) while watching podcasts, being able to repeat and rewind certain parts of podcasts, being able to watch a podcast more than once from any location, a new dimension of the lecture emerges: "The pedagogical deployment of the podcast within the university can be characterized as being in a transitional phase—slipping between different functions and usages: as lecture supplement, as partial replacement, and, increasingly, as a simulacrum of the lecture itself" (143). As noted above, the image of the lecture embodied in the charismatic professor of the literature classroom compels film and television plots, but McKay and Brass contend that there is something of the lecture (and, for them, it is the field of cultural studies, which largely came from English departments) that cannot be "mechanically reproduced" and thought to be as effective as an in-person lecture (156). In the humanities they note a loss of "authenticity" from "the aura of the lecture" (155). Moreover, this "authenticity" that McKay and Brass discuss in the cultural studies classroom is one that they tie to the necessary timeliness of their discipline; however, the fact remains that in a replicated lecture preserved in video, it is one version of that lecture repeated and frozen. The improvisational element that speakers bring, sensing moods in the room, watching students responses, or being interrupted with questions all become frozen an unable to change unless the lecture is redone. Understanding the social connectedness of the students in the room and the changes that a lecture takes from semester to semester when new historical and cultural waves recontextualize received knowledge is part of what makes students inclined to favor attending lectures in person. So, while students are shown to wear "the badge of efficiency," as McKay and Brass borrow from Thorstein Veblen's *The Theory of the Leisure Class* (1912), by sitting attentively with their lecture-viewing equipment, they have mixed responses to this approach.

While the interaction (in the form of repetition on demand) that the podcast or TED Talk promises beckons students to participate, albeit in a limited way, the 1953 issue of *College English* that dedicated itself to asking if the lecture should be used "for background only" also cues the same concern: Does the lecture (digitized or in person) form demand passivity of students? The mid-twentieth-century concern as well as current ones centers themselves on the large lecture, which shows that the form of knowledge transfer takes a shape that is driven by maximizing tuition dollars. The fate of the lecture seems not to be in jeopardy any more than it had been in literary studies in the mid-twentieth century, or, even as it is said to have been "destroyed" by Harvard University's replacing the trivium, especially rhetoric, with literary studies (Heinrichs 37).[10]

Long removed from Matthew Arnold's "The Function of Criticism at the Present Time" (1865), which advanced the place of literature in culture as the source to locate "the best that is known and thought in the world," is the state of literary study in the twentieth century.[11] For Arnold, the study of literature could and should exist outside of the world of history. However, within his essay about the function of criticism, which has largely been understood to mean the endpoint of the interpretation of literature, there is an interest in the process of criticism. Ironically, such an interest strikes at the center of what Carens identifies as being the sublimated discomfort that students feel when they are being asked to take the role of the authority and produce meaning themselves. When Arnold explains that "the creative faculty" and the "critical faculty" are basically the same, he is not just making a rhetorical move to raise criticism to the high esteem that poetry enjoys but also making an observation that the act of interpretation is a skill that benefits the practitioner. The lecture in the twentieth century in my class and my colleagues' literature classrooms, which are situated in a state regional baccalaureate university with class sizes of thirty-five and under, remains an effective touchstone for covering the historical background of authors, setting up received critical points regarding a text, and/or establishing a context for the discussion of the day. However, cultivating this other aim of criticism, not at the wisdom that an interpretation may find at the end of the critical exercise but rather at the process of interpretation itself, the uneasy building of a reading that always can take a different turn and yield a different result, may give truth to insecurities, complexities, self-evaluation, and, finally, autonomy that popular culture delivers in the way of romantic relationships.

To the point that "the lecture" in literary studies is a myth that exists in the imagination of popular media, like all myths, there is a truth that is more felt than factual. In this way, the lecture endures while its function as an avatar of learning wanes in a practical sense. More importantly, the broader context in which people learn in the twenty-first century has delivered a blow to the humanities and to the literature major. More often than not, those who practice literary interpretation are asked to explain how well literature majors fare in law school,[12] how they will be prepared for a variety of graduate-level tracks, and how their degrees are flexible and capable of delivering them into a variety of business sectors. Departments of English laude the usefulness of analytical skills, boosted vocabularies, cultural literacy, and writing skills as they compete with professional departments in the health sciences, hard sciences, business schools, and other majors that are believed to deliver not just a degree but a rewarding life. The waning of numbers of English majors (dropping 20% from 2012 to 2016)[13] has meant that there are fewer tenure-track jobs to teach the next class choosing English or literary studies for graduate training.

However, the lecture in the literature classroom can extend beyond it and continue to play a role in entirely new territories. Displaced by a lack of demand, greater dependence on precarious faculty positions, and online learning in a variety of venues that lie outside of the traditional university, scholars trained in literary studies retain the process of interpretation, the strange hybrid that at once tells what something means and begs for alternatives, have launched a variety of places where they can "lecture," or share knowledge and the results of their interpretations while also making a living. Public scholarship, such as that practiced by Hannah McGregor, and Marcelle Kosman, invested their time into what has become a successful podcast, *Witch, Please*, in which they analyze all things in the Harry Potter industry through a feminist lens.[14] In the podcast, something that can be consumed passively, like a lecture, they pontificate and explain. In social media and elsewhere their audience, such as students in a classroom, respond, disagree, and offer variations, more interpretations, corrections, and a lively conversation. She says that she enjoys the audience's responses: "Much has been made of Internet call-out culture, but appeals for accountability that come from within a community are a different thing altogether. That kind of critical commentary opens up dialogue and creates new, shared vocabularies. It builds, instead of shutting down" (5). Even without the power dynamic of grades, discussions about texts call for investment by participants.

The lecture has never been a fixed entity. Its appearance as such is an ahistorical manifestation of something important, however. It serves the purpose of ameliorating the awkward and vulnerable feelings that doing qualitative analysis brings forth. As in a dream, codifying a frightening experience in terms that are familiar, the shared dream of film and novels make the professor the embodiment of the knowledge; he delivers knowledge in a lecture, and his human failures give way to an interpersonal drama. Of course, as Carens points out, the real drama happens in us, as we share our interpretations of what strikes us or interests us. The lecture is but one part in a circuit of making meaning and in gathering the skills and understanding necessary to live in a complex, diverse, accelerated world.

APPENDIX

In a course, called Writing Interpretive Papers, junior-level students are encouraged to leave our class which hybridizes lecture, discussion, and writing workshops in order to build their own independent discussions online. Below is the assignment sheet, including sample prompts, and the grading rubric.

ONLINE GROUP PROJECT FAQ

Why Do We Do Group Work?

Local reason: Group work is important in this class, because the readings are difficult, and the terms and ideas are quite new to most students. The project keeps students in practice, using these difficult terms and ideas during the week, so as to gain better control over them.

Global reason: The transferable skills that students gain coordinating time and communication with other people make them successful in a variety of workplace settings. In the case of being able to use difficult language, abstract ideas, and logical reasoning, students are positively positioned in high-level and highly competitive working environments.

Why Must I Post Twice Each Week?

Students can begin to feel removed from class or, worse, forget about it until hours before class, reading at the last minute; the material is just too difficult for that habit.

If I skip a couple of weeks, can I make it up by posting a lot on one day?

No. Posts are expected to come in on a regular basis to keep you in touch with course content.

What Should I Post?

Questions about terms, ideas, and passages are a great place to start, especially in this class where the readings may alienate you at first. It is unlike any other literature class, because it addresses the ideological and cultural environment in which we read. Reading right before class begins is not helpful, because the readings are dense and complicated. Posting insights and observations is helpful, too. If you notice, a key idea, such as hybridity, and it comes up in a film you have seen or a text you are reading for another class, this forum is a great place to make the connection.

These are not full papers. They should present a solid idea or response to another's idea in about a paragraph.

If you are still feeling lost, you may want to consider your posts to act like journal entries that others can read and respond to. There is also a sample group project for you to view on Canvas.

If a group member does not keep up with posts or the posts are not good, will I get a lower grade?

No, grades are determined individually, based on the attached rubric.

Does spelling count?

Spelling and grammar are not my priorities as I read the posts, though if the problems are so severe so as to obscure meaning, then they will negatively affect the score of the individual.

How am I graded?

Please note that scores are assigned individually as this rubric will reflect the score of each group member; there will be no collective group grade for this assignment. *Failure to post for a week (seven consecutive days) or longer will result in a severe penalty.*

Sample prompts:

1. In this first week's discussion, share your personal experiences regarding the interpretation of literature vis–à–vis (face-to-face with) the way that Tyson explains New Criticism. To begin, you can describe how interpretation has been taught in other courses, and once you have read Tyson, post again to discuss how your experience is similar to or different from what the New Critics, also known as "formalists," do.
2. Without necessarily taking sides or being judgmental, consider the ways in which a purely formalist approach opens up insights and meanings (in Brooks's essay) and creates blind spots for critics, such as in Bloom's case. Discuss your own comfort or discomfort with the idea that we can appreciate literature without bringing "politics" into it.

Table 4.1 Online Posting Rubric*

Keeping up with Posts					
Posts are made in a timely way	(Not kept throughout) 1–4	5–8	9–13	14–16	(Kept throughout) 17–20
Threads are logical	(8–10 posts are short) 1–2	3–4	5–6	7–8	(1–2 posts are short) 9–10
Members participate equally	(Unequal participation) 1–2	3–4	5–6	7–8	(Equal participation of all members) 9–10
Quality of Posts					
Ability to use ideas	(Reiterates ideas) 1–2	3–4	5–6	7–8	(Shows grasp of ideas) 9–10
Use of new terms	(Uncontrolled/ incorrect) 1–2	3–4	5–6	7–8	(Accurate and specific) 9–10
Quotes	(Stand-alone) 1–2	3–4	5–6	7–8	(Interpreted for reader) 9–10
Examples	(Unconnected to class's ideas) 1–2	3–4	5–6	7–8	(Clearly connected to theories) 9–10
Explanation	(Repetition/ redundancy) 1–2	3–4	5–6	7–8	(Creates insights) 9–10
Approach to Assignment					
Intellectualism	(Lacking curiosity) 1–2	3–4	5–6	7–8	(Engaged, curious) 9–10

*Please note that scores are assigned individually as this rubric will reflect the score of each group member, and there will be no collective group grade for this assignment. *Failure to post for a week (seven consecutive days) or longer will result in a severe penalty.*

NOTES

1. Jean Francois Lyotard, *The Postmodern Condition* (Manchester: Manchester University Press, 1979).
2. Sterg O'Dell, "Lecture for Background Only." *College English* 17/2 (1955): 105–6.
3. Monroe Beardsley and W. K. Wimsatt, "The Intentional Fallacy." In *Literary Theory: An Anthology*, Julie Rivkin and Michael Ryan, eds. (Hoboken: Wiley Blackwell, 2017), 29–41.
4. Linda Adler-Krassner, "Liberal Learning, Professional Training, and Disciplinarity in the Age of Educational 'Reform': Remodeling General Education." *College English* 76/5 (2014): 436–57.

5. Timothy L. Carens, "Serpents in the Garden: English Professors in Contemporary Television and Film." *College English* 73/1 (2010): 9–27.

6. Carens cites films as early as the 1970 *Looking for Mr. Goodbar* to the 2008 film *Smart People*. A host of other titles are treated as well, including *Animal House* (1978), *A Change of Seasons* (1980), *Terms of Endearment* (1983), *D.O.A.* (1988), *Mr. Wonderful* (1993), *One True Thing* (1998), *Loser* (2000), *Miss Congeniality* (2000), *Wonder Boys* (2000), *The Rules of Attraction* (2002), *The Squid and the Whale*, (2005), and *Elegy* (2008).

7. Cleanth Brooks, "The Formalist Critics." In *Literary Theory: An Anthology*, Julie Rivkin and Michael Ryan, eds. (Hoboken: Wiley Blackwell, 2017), 15–20.

8. Matthew Arnold, "Dover Beach." *The Poetry Foundation*. Available at http://www.poetryfoundation.org/bio/matthew-arnold.

9. Kieryn McKay and Kylie Brass, "Hired Hands: Casualised Technology and Labour in the Teaching of Cultural Studies." *Cultural Studies Review* 17/2 (2011): 140–64.

10. Jay Heinrichs, "How Harvard Destroyed Rhetoric." *Harvard Magazine* 97/6 (1995): 37–42.

11. M. Arnold, S. Collini, and M. Arnold, *Culture and Anarchy and Other Writings* (Cambridge: Cambridge University Press, 1999).

12. Denise Forster, "Using Literature to Make Better Lawyers." *UVA Lawyer* (2005). Available at https://www.law.virginia.edu/static/uvalawyer/html/alumni/uvalawyer/f05/literature.htm.

13. Colleen Flaherty, "The Evolving English Major." *Inside Higher Ed* (July 18, 2018). Available at https://www.insidehighered.com/news/2018/07/18/new-analysis-english-departments-says-numbers-majors-are-way-down-2012-its-not-death.

14. H. McGregor, "Podcasting, Public Scholarship, and Accountability." *MLA Newsletter* 50 (2018): 4–5.

WORKS CITED

Adler-Krassner, Linda. 2014. "Liberal Learning, Professional Training, and Disciplinarity in the Age of Educational 'Reform': Remodeling General Education." *College English* 76/5: 436–57.

Arnold, Matthew. n.d. "Dover Beach." *The Poetry Foundation*. Available at http://www.poetryfoundation.org/bio/matthew-arnold.

Arnold, M., Collini, S., and Arnold, M. 1999. *Culture and Anarchy and Other Writings*. Cambridge: Cambridge University Press.

Beardsley, Monroe and W. K. Wimsatt. 2017. "The Intentional Fallacy." In *Literary Theory: An Anthology*, Julie Rivkin and Michael Ryan, eds. Hoboken: Wiley Blackwell. 29–41.

Brooks, Cleanth. "The Formalist Critics." In *Literary Theory: An Anthology*, Julie Rivkin and Michael Ryan, eds. Hoboken: Wiley Blackwell. 15–20.

Carens, Timothy L. 2010. "Serpents in the Garden: English Professors in Contemporary Television and Film." *College English* 73/1: 9–27.

Flaherty, Colleen. 2018. "The Evolving English Major." *Inside Higher Ed.* July 18. Available at https://www.insidehighered.com/news/2018/07/18/new-analysis-english-departments-says-numbers-majors-are-way-down-2012-its-not-death.

Forster, Denise. 2005. "Using Literature to Make Better Lawyers." *UVA Lawyer.* Available at https://www.law.virginia.edu/static/uvalawyer/html/alumni/uvalawyer/f05/literature.htm.

Heinrichs, Jay. 1995. "How Harvard Destroyed Rhetoric." *Harvard Magazine* 97/6: 37–42.

Lyotard, Jean Francois. 1979. *The Postmodern Condition.* Manchester: Manchester University Press.

McGregor, H. 2018. "Podcasting, Public Scholarship, and Accountability." *MLA Newsletter* 50 (December): 4–5.

McKay, Kieryn and Kylie Brass. 2011. "Hired Hands: Casualised Technology and Labour in the Teaching of Cultural Studies." *Cultural Studies Review* 17/2: 140–64.

O'Dell, Sterg. 1955. "Lecture for Background Only." *College English* 17/2: 105–6.

Chapter 5

Discourse on Lecture Comprehension in the Twenty-First-Century Classroom

Teaching with an Awareness of the Cultural Construction of Language

Monika Dix

Teaching a foreign language involves instructors spending a lot of time with students helping them to learn to express themselves in their non-native language through proper usage of vocabulary, grammar, and culture. In this context, the traditional face-to-face in-person lecture can serve as an effective teaching method through which instructors can advance students toward fluency by sharing expert knowledge, personal experiences, introducing new customs, and teaching discursive strategies, with the aim that students will reach a level of linguistic and cultural competency where they can fully and appropriately express their own thoughts. However, especially with the growing demand of online classes to accommodate twenty-first-century learners, the central role of the traditional lecture has been largely underestimated and the critical scholarship on this topic has been lacking. This presents the questionable view that the traditional lecture, with its face-to-face in-person interaction between instructors and students, constitutes a peripheral topic regarding the discourse on lecture comprehension. The discussion about the effectiveness of the lecture in foreign language instruction has rarely merited the same rigorous analysis—for example, electronic media instruction such as online classes, which have gradually become an established part of twenty-first-century education.

As a multilingual teacher and translator, I find the traditional face-to-face in-person lecture an effective approach to teaching foreign languages. Since teaching foreign languages with an awareness of the cultural construction of language is a key aspect in the discourse on lecture comprehension in the twenty-first-century classroom, I have spent a large part of my teaching career exploring the terrain between English and Japanese. In particular, my focus has been on reconsidering and utilizing the spaces between these two languages as an effective pedagogical tool to assess lecture comprehension. In contrast to the common perception that the terrain between two very different languages, such as English and Japanese, constitutes a kind of no-man's land that simply needs to be traversed as quickly as possible in order to advance students toward fluency in the language they are studying, we might think about it as a more productive space in which ideas flow together and new dynamic, creative forms of cultural expression might emerge. In this chapter, I will argue for the value and importance of those types of expressions that arise in the vast transitional spaces between languages. My primary focus is on listen-to-summarize tasks that represent listeners' meaning building and the discourse construction of the lecture for listening assessment purposes. By advocating the theoretical construct of the 3Ps—namely products, practices, and perspectives—combined with an inquiry-based teaching approach, this study discusses the nature of lecture comprehension. In addition to promoting student motivation and engagement that can help overcome past issues of stereotyping and lack of intercultural awareness, it also highlights teachers' and students' perceptions of listen-to-summarize task demands and their respective implications. It is my goal to illustrate how teachers can enhance the traditional lecture with inquiry learning into their lectures to create a rich and meaningful environment in which students interact with authentic material in the target language and fill the transitional spaces between languages with their own understanding of foreign cultures' products, practices, and perspectives.

There is common consensus among educators that language and culture are closely linked and should be acquired together.[1] Although there has been a considerable increase in classroom teaching time on culture in foreign language instruction over the past twenty-five years, there is still much debate about how to develop and teach a lecture with an awareness of the cultural construction of language.[2] In other words, what role does culture play in foreign language teaching and how is culture teaching successfully integrated into the traditional classroom lecture?

Based on my own personal experiences of teaching Japanese language in the traditional face-to-face in-person lecture style, I argue that Japanese language and culture learning modules can be used in the classroom to optimize students' "out of the box" thinking. These modules are content-based instructional units, which focus on culture oriented-learning in accordance with the ACTFL (American Council of Teachers of Foreign Languages) proficiency levels.[3] The research questions that guided this study are as follows:

(1) What is culture?
(2) What are the main considerations for designing and teaching a lecture that successfully integrates Japanese language and culture?

(3) For students of Japanese, what are the main factors that influence the instructor's choice of subject matter?
(4) How can students gain real-world competency in the classroom?

The definition of "culture" is a problem in itself due to the broad concept of the term. Culture embraces many aspects of human life, such as customs, beliefs, social forms, and material traits of a racial, religious, or social group, as well as characteristic features of everyday existence shared by people in a place or time.[4] Especially for foreign language educators, this multifarious composition of "culture"—lacking a clear definition—makes it difficult to decide which aspects of the target culture should be taught in the lecture. Regarding the definition of "culture" that I am referring to in this study it is one that includes philosophical perspectives, behavioral practices, and cultural products of a society. These cultural products might be tangible, such as for example art or literature, or intangible, such as for example political systems. The Japanese language and culture learning modules I use in my lectures contextualize classroom activities and relate them to real life situations. By addressing this kind of inquiry learning, my lectures create a rich and meaningful environment in which students interact with authentic material in the target language and bridge the "no-man's land" between languages with their own understanding of foreign cultures' products, practices, and perspectives.

According to the Standards of Foreign Language Learning, the presently available most effective methods for teaching culture and achieving cultural proficiency are as follows: (1) students should demonstrate an understanding of the relationship between the practices and perspectives of the culture studied, and (2) students should demonstrate an understanding of the relationship between the products and perspectives of the culture studied.[5] By advocating the theoretical construct of the 3Ps—products, practices, and perspectives—combined with an inquiry-based teaching approach, my lectures explore culture in a systematic and contextual way, while at the same time leaving room for some flexibility. In order to show how this kind of framework enforces the listen-to-summarize tasks that represent listeners' meaning building and the discourse construction of the lecture for listening assessment purposes, I will give an example from my 300-level advanced Japanese conversation class.

Saginaw Valley State University is a small, public, and teaching-intensive university in the state of Michigan where I teach in the Department of Modern Foreign Languages. In 2010, I started the Japanese language program, which has been gradually growing, and being the only instructor I have to teach all levels of Japanese. Although the majority of students who minor in Japanese at Saginaw Valley State University are International Studies majors, especially in the past two years, the number of students who major in STEM subjects and minor in Japanese has noticeably increased. This prompted me to come up with an innovative design thinking for my upper-level advanced Japanese conversation class (JAPN 332) in order to give students the opportunity to learn about real-world competency for their future careers through the integration of Japanese and their majors, especially STEM. The goal of this project is to design content-based instructional modules for hands on, subject matter and culture oriented-learning in accordance with the ACTFL proficiency levels. Focusing on vocabulary-learners resources for proficiency competence in their academic majors, the rationale is that students with intermediate language proficiency (usually obtained after studying two years of a foreign language at a university) also acquire basic knowledge of the subject of their major (biology, chemistry, physics, math, etc.) so that they can interact academically and socially with other Japanese language learners who share the same academic interests (internships, conferences, etc.). Lastly, Japanese and subject learning modules can improve students' career-path knowledge by offering Japanese language activities related to employment. The outcomes are to enhance knowledge of Japanese contributing to STEM, make Japanese attractive to new learners, and cultivate a broader set of skills, especially cultural proficiency.

Prior to the beginning of the 2016 fall semester, I contacted all students who had enrolled in JAPN 332 and asked them to submit the following information:

(1) their major
(2) their specialization in their major
(3) their minor

The results were as follows: out of thirteen students, two majored in international studies; two in art; two in communication; one in creative writing; one in criminal justice; one in geography; one in business; one in physics; one in math; and one in biology. They all minored in Japanese.

Based on this information, I designed a vocabulary-learning resource for proficiency competence in the students' academic major subjects. The collection of data in this study was obtained through weekly class discussions and student-teacher feedback on the progress report, as well as the instructor using a structured observation protocol and informal conversational interviews with the students throughout the 2016 fall semester. Data was analyzed and assessed by using the ACTFL Intermediate Interpersonal Mode Rubric as shown in[6] table 5.1.

Since this was the first time I exposed my 300-level students to such a project, I used scaffolding. First, I ask students to make a list of ten keywords in English and in Japanese they consider essential for a basic conversation about

Table 5.1 Meets Expectations

Category	*Exceeds Expectations*	*Strong*	*Weak*	*Does Not Meet Expectations*
Language Function Language task student is able to handle in a consistent, comfortable, spontaneous manner	language expands narration that includes connectedness and cohesiveness	CREATES with language; ability to express own meaning expands in quantity and quality	CREATES with language; ability to express own meaning in a basic way	mostly memorize language with some attempts to create
Text Type Quantity and organization of language discourse; coherent paragraphs	paragraph-like discourse	advanced sentence-level discourse; complexity	simple sentences	memorized phrases
Communication Strategies Quality of engagement and interactivity; how student participates in conversation and advances it	conversation using a variety of strategies	maintains conversation by asking and answering questions	simple conversation asks/answers simple questions	responds to basic questions
Clarification Strategies How student handles breakdown of comprehension; what one does when one partner does not understand the other	clarifies by paraphrasing	clarifies by asking and answering questions	clarifies by asking questions	clarifies by selecting substitute words
Comprehensibility Understanding student's meaning (teacher vs. native speaker); How independent of teaching situation is this conversation?	generally understood by those unaccustomed to language learners	generally understood by those accustomed to language learners	sometimes understood	rarely understood
Language Control Accuracy, style, vocabulary, fluency	accurate in everything	mostly accurate	sometimes accurate	only accurate with memorized phrases

their major. During the class lecture, students had to complete various listen-to-summarize tasks both in English and in Japanese. Through this exercise, which is most effectively executed in a face-to-face lecture, students became aware of the different cultural construction of English and Japanese. Unlike the common stereotypical assumption that English words can just be translated into Japanese based on their meaning, students learned to build meaning based on the understanding of foreign cultures' products, practices, and perspectives. This discourse construction of the lecture, which uses teaching strategies for cultural awareness including the comparison method, provided me, the instructor, with a new module for listening assessment purposes. The examples are as follows: (table 5.2).

The next step was for students to develop a bilingual English-Japanese glossary entry, which they then had to further develop and expand into the final presentation. In order to determine how effective students were in achieving both the linguistic and cultural competency, the lecture focused on the discussion of the relationship between products, practices, and perspectives. For example, after several drafts and feedback from the instructor, the student majoring in biology and specializing in genetics provided the following:

Biology—生物学　（せいぶつがく）

gene mapping—遺伝子地図（いでんしちず）

Throughout my JAPN 332 lectures, I observed a gradual shift from teaching to learning. The acquisition of language and culture transitioned from teacher lectures to students discovering culture first hand through inquiry learning, placing the students and their projects at the center of the learning process. At this step students then had to come up with (1) additional fields of expertise which use the term "gene mapping" such as biotechnology, biochemistry, and microbiology; (2) an English and Japanese definition of gene mapping; and (3) English and Japanese sample sentences which include gene mapping. The English-Japanese entries (2) and (3) can be, but do not have to be, translations of each other. This was the result:

Table 5.2 Biology

Biology	生物学	(せいぶつがく)
Anatomy	解剖学	(かいぼうがく)
Biochemistry	生化学	(せいかがく)
Gene mutation	遺伝子突然変異	(いでんしとつぜんへんい)
Genetics	遺伝学	(いでんがく)
Genotype	遺伝子型	(いでんしがた)
Incubator	孵卵器	(ふらんき)
Petri dish	ペトリ皿	(ペトリさら)

Stem Subject

Biology—生物学（せいぶつがく）

Specialty

gene mapping—遺伝子地図（いでんしちず）

Additional Fields of Expertise

biology (-tech, -chem, micro-)

Definition

Determination of relative positions of genes on a DNA molecule and of the distance, in linkage units or physical units, between them.

遺伝子地図とは染色体上の遺伝子の位置を示した地図のこと。

Sample Sentences

Gene mapping of the genomes of fishes will only serve as the starting point for more detailed studies into the function of these genomes. Future research for understanding the mechanisms of gene actions and interactions will benefit areas of biology, ecology, ethology, evolution, and physiology.

遺伝子地図とは染色体上の遺伝子の位置を示した地図のこと。金銭的な問題や効率面からゲノムプロジェクトの前段階として各種の地図が作られる地、実際の研究では塩基配列全ての情報よりも、こういった簡略化された地図が有用な場合も多いです。

This paradigm shift from passive receiver of information to active builders of knowledge, linguistic and cultural proficiency engaged students in effective learning that promotes higher-order thinking, exploring the hidden meaning in the transitional spaces between languages.

Next, I had to come up with a way for students to gain real-world competency in Japanese and their academic subject matter by applying the glossary they created. An essential content of JAPN 332 is business Japanese. By combining business Japanese, including language, behavior, social etiquette, with the Japanese plus culture modules tailored to each student's major and area of specialization, this innovative course design optimizes student's "out of the box" thinking. Ideally, I would have liked my students to visit local companies to apply this acquired knowledge, but due to geographical and time restraints this was not possible. Instead, I invited our Japanese exchange professor, whose specialty is computer and information science, to our class a couple of times. Between the two of us, we divided the class into humanities and science majors. Then we held mock interviews in which students had to sell themselves on the job market using the Japanese + culture module designed in class.

Overall, the project was very well perceived by both instructor and students. Students especially enjoyed applying their Japanese language skills with their STEM knowledge of their major, something they have never done before. In terms of improving students' Japanese language skills, the Intermediate Interpersonal Mode Rubric is a successful assessment tool. Out of the thirteen students enrolled in JAPN 332 and who participated in this study, two students exceeded expectations in the categories of language function, text type, communication strategies, clarification strategies, comprehensibility, and language control. Three students strongly met the expectations, whereas four students weakly met the expectations. Four other students did not meet the expectations.

There are several implications from this study for students, instructors, administrators, and schools/universities for considering implementing STEM + languages to counteract the decline in foreign language studies and to recruit more students to the study of Japanese. Over time, employing a student-centered approach on a 300 level Japanese language course with well-structured activities allows for instructors to be more comfortable with curriculum adjustments/expansions in new ways for twenty-first-century learners and for students to be successful.

As this chapter has shown, project-based learning (PBL) in a classroom fosters foreign language learning by engaging students in important, real-world projects and having students utilize their first and second language knowledge—in the case of my JAPN 332 students at Saginaw Valley State University English and Japanese—skills and abilities to complete the work. Project-based learning integrates all four language skills (listening, reading, speaking, and writing), incorporates collaborative team work, lends itself to clear outcomes, and challenges learners to use the foreign language in different capacities and contexts.

Underlying my innovative design thinking for twenty-first-century learners to develop real-world competencies through the integration of Japanese + STEM in the classroom lecture were the following components:

1. Teaching:
    (a) Lesson Planning
        focus on connections, translations, understand student misconceptions, understand student capabilities, problem solving, focus on big ideas, real-world and cultural relevance
    (b) Classroom Practice
        justify thinking, writing for reflection, cooperative learning, inquiry
2. Efficacy: content and pedagogical knowledge result in positive self-efficacy, commitment, and planning is crucial
3. Materials: linguistic and technological resources, instructor support

My major goal in creating the inquiry questions for my lecture that provoke interest in the target language and culture was to develop students' understanding of cultural differences. Therefore, I suggest that the traditional face-to-face in-person lecture is an effective mean of instruction because it draws on the spontaneous human aptitude for communication. In addition to promoting student motivation and engagement that can help overcome past issues of stereotyping and lack of intercultural awareness, it also highlights teachers' and students' perceptions of listen-to-summarize task demands and their respective implications. I have illustrated how teachers can enhance the traditional lecture with inquiry learning into their lectures to create a rich and meaningful environment in which students interact with authentic material in the target language and fill the transitional spaces between languages with their own understanding of foreign cultures' products, practices, and perspectives.

NOTES

1. H.D. Brown, *Principles of Language Learning and Teaching* (New York: Pearson Education, 2007).
2. See R.C. Lafayette, "Integrating the Teaching of Culture into the Foreign Language Classroom." In *Toward a New Integration of Language and Culture*, A.J. Singerman, ed. (Middlebury, VT: Northeast Conference on the Teaching of Foreign Languages, 1988), 47–62; Z. Moore, "Technology and Teaching Culture: What Spanish Teachers Do." *Foreign Language Annals* 39/4 (2006): 579–94.
3. Jacque Bott Van Houten & Kathleen Shelton, "From Culture to Intercultural Communication: Framing the Cultural Landscape of the Classroom with CAN DO STATEMENTS." *The Language Educator* 13/1 (2018): 34–39.
4. H. N. Seelye, *Teaching Culture: Strategies for Intercultural Communication* (Lincolnwood: National Textbook Company, 1993), 15.
5. National Standards in Foreign Language Education Project, *Standards for Foreign Language Learning in the 21st Century* (New York: Author, 1999).
6. ACTFL Intermediate Interpersonal Mode Rubric, Available at https://education.ohio.gov/getattachment/Topics/Ohio-s-New-Learning-Standards/Foreign-Language/World-Languages-Model-Curriculum/World-Languages-Model-Curriculum-Framework/Instructional-Strategies/Assessment-Guidance-and-Sample-Rubrics/IPA-AppendixF_Rubrics-ACTFL.pdf.aspx.

WORKS CITED

ACTFL Intermediate Interpersonal Mode Rubric. n.d. Available at https://education.ohio.gov/getattachment/Topics/Ohio-s-New-Learning-Standards/Foreign-Language/World-Languages-Model-Curriculum/World-Languages-Model-Curriculum-Framework/Instructional-Strategies/Assessment-Guidance-and-Sample-Rubrics/IPA-AppendixF_Rubrics-ACTFL.pdf.aspx.

Brown, H.D. 2007. *Principles of Language Learning and Teaching*. New York: Pearson Education.

Houten, Jacque Bott Van & Kathleen Shelton. 2018. "From Culture to Intercultural Communication: Framing the Cultural Landscape of the Classroom with CAN DO STATEMENTS." *The Language Educator* 13/1: 34–39.

Lafayette, R.C. 1988. "Integrating the Teaching of Culture into the Foreign Language Classroom." In *Toward a New Integration of Language and Culture*, A.J. Singerman, ed. Middlebury, VT: Northeast Conference on the Teaching of Foreign Languages. 47–62.

Moore, Z. 2006. "Technology and Teaching Culture: What Spanish Teachers Do." *Foreign Language Annals* 39/4: 579–94.

National Standards in Foreign Language Education Project. 1999. *Standards for Foreign Language Learning in the 21st Century*. New York: Author.

Seelye, H.N. 1993. *Teaching Culture: Strategies for Intercultural Communication*. Lincolnwood: National Textbook Company.

Chapter 6

The Continuing Value of Lecture in History Education

Hyrum Lewis

For millennia, history educators have been lecturing to students; however, for the past few decades, critics have been arguing that lecture bores students and that we should throw out this antiquated method in favor of something new and more exciting. Furthermore, many claim that we can just "put lectures online" and spend precious in-class time on more valuable activities than "passive" instruction.[1]

The following chapter defends the value of in-person lecture as an important means to teach historical knowledge and skills. Drawing upon recent research in pedagogy and cognitive psychology, I will argue that the traditional lecture method has been used to teach history for thousands of years for a good reason: it works.[2] Historians would be wise to continue to use the lecture method to instruct students as long as certain conditions are met.

The prejudice against lecture (i.e., "expert instruction") stems primarily from its antiquity. Historians have been lecturing at least since Herodotus in the fifth century BCE. The common refrain from opponents of expert instruction is that since historians have been lecturing since ancient times, it's obviously outdated. Anyone who wants to be "fresh," "innovative," and "progressive" should naturally be moving past this hoary teaching method.[3]

But notice that we have also been drinking water, breathing, exercising, and eating vegetables since ancient times, yet few people admonish us to stop these practices. Since age alone is not a sufficient reason to abandon a practice, opponents of expert instruction must find a different rationale on which to criticize lecture. In fact, one might just as easily argue that the longevity of expert instruction is an argument in its favor: history professors have been lecturing since the time of Herodotus because, like exercising, it has enduring value. In a Darwinian sense, the lecture method would not have survived as the dominant model over the centuries were it not effective.

One charge against expert instruction is that it is a form of "passive" rather than "active" learning.[4] Again, this is an unfounded prejudice. The activeness of instruction depends upon the approach and skills of the professor and the disposition of the students. "Active learning" versus "expert instruction" is a false dichotomy since lectures *can be extremely active*.[5] Active learning does not suggest a particular behavior or method, but a mindset and level of engagement in which students "take control over their own learning."[6] Some of the most active learning moments in many students' lives occur when they are deeply engaged in listening to a spellbinding historian lecture about a topic in which they have expertise and passion.[7]

Even the most vocal anti-lecture zealots have admitted to me in private that abandoning expert instruction usually results in a drop in student learning. But instead of taking this as evidence that lecture might have value, they assume that the teacher must not be implementing new learning activities correctly. The confirmation bias is stunning: if a lecture is boring, critics tell us that the problem is with the lecture method itself. If an alternative method is boring and produces poor learning, critics conclude that the teaching method is right, but just being poorly implemented. The reality is that different methods of teaching history will be effective depending upon the skills and efforts of the instructor. Most history teachers have a gift for lecturing (it is probably a main reason they went into the field in the first place) and demanding that they stop the practice would deprive students of valuable learning opportunities.[8]

In reality, the lecture, when done correctly and under the right conditions, has been shown to be a remarkably effective method for teaching history.[9] In particular, I argue that in-person lecture has two major advantages over other methods of teaching history (including online lectures): (1) lecture instills enthusiasm and intrinsic motivation in students via positive modeling, and (2) lecture promotes "level 2" expertise.

History lecturers who exemplify certain traits are more likely to see those traits develop in those they instruct. For most students, their teachers don't just know the subject under study, they *are* the subject under study. Students form nearly their entire impression of the field of history from the teacher. An excited teacher lecturing on a subject with passion has a huge impact on student love of the topic.[10] For little-understood reasons, the value of this modeling is almost entirely lost in online formats.[11]

Enthusiasm for a subject is highly correlated with intrinsic motivation, the growth mindset, and grit. History students often develop these traits by seeing them modeled by their professors. A passion for history and a willingness to work hard, sacrifice, and develop historical skills is often transferred from teacher to student through lecture.[12] The difference between a strong student or a poor student is usually explained by motivation rather than intelligence.

Excellent history students are not necessarily those with the most inborn cognitive ability, but those with the most love of the subject (which then, through the effort that comes from passion, translates into cognitive ability). The best lectures involve emotions that push deep processing and create knowledge relationships. The future of the historical profession depends upon developing young, talented students who have the drive, tenacity, and self-direction to finish a graduate program, spend the requisite time in the archives, and do the painstaking work to produce high-quality historical research.[13] These traits are often instilled during historical lectures. A student's love of history nearly always comes from a teacher who modeled their own love of history in lecture.

Most importantly, expert instruction is one of the best ways to ignite the "intrinsic motivation" to learn history. An intrinsic desire to learn a subject results in better long-term outcomes than "extrinsic motivation"—the promise of a high grade or other external reward. One recent study has split interest into two distinct types: (1) individual interest, which is internal and lasting, and (2) environmental interest, which is external and fleeting.[14] Individual interest corresponds and leads to intrinsic motivation while extrinsic motivation can actually harm long-term student learning.[15] Intrinsic motivation is crucial to deep and sustained learning, and yet it's not just there, it must be instilled. Expert instruction, as long as it is co-spatial, has been shown to be among the most effective ways to instill this intrinsic motivation. Excitement and enthusiasm rub off on students and come from clear, captivating explanations from great teachers.[16]

Much of this value is lost in online lecture formats. The classroom, unlike the online lecture, is a site of interaction, not just a site of one-way transmission.[17] In a proper lecture environment, students are "working with" teachers rather than "receiving from" them. There is a deep connection between setting, motivation, and learning, but online education cannot yet replicate this.[18] The classroom dynamic leads to community which in turns leads to excitement and motivation. In-person instruction not only helps students learn at a higher level but also helps them have higher performance in subsequent courses, even in unrelated fields.[19]

Video lectures can actually harm learning by distracting the student. A conversational instruction with audio and visual elements in a face-to-face setting has been shown to be one of the best learning tools there is while online lectures usually lack the simultaneous visual and auditory full experience of face-to-face instruction and are therefore less effective.[20] "Connecting with or relating to a fairly specific audience is central; and the most motivating audience seems to be a face-to-face audience."[21]

Expert instruction also takes advantage of the principle of shared attention. Learning is enhanced when there is a community of learners jointly focused on a common subject.[22] Not only does face-to-face expert instruction enhance historical learning, but it also promotes socialization and helps build the sense of community and belonging that are a crucial part of any education. Inasmuch as modeling-through-lecture promotes that sense of self-direction, then lecture in history is a powerful and important teaching method and the future of the profession largely depends upon it.

Students learn history better not only when they are excited about the subject, but when they are excited about the period. The presence of other people generates this excitement in ways that don't happen in a video or telecommunicated interaction.[23] According to student surveys, the most important characteristic of great teachers is "exciting." This was even more important than "knowledgeable," "accessible," and "engaging," but each of those traits is related to excitement and is greatly reduced when not face-to-face.[24]

Since motivation is among the most important factors contributing to long-term student learning and success, the value of face-to-face lecture is hard to overstate. As two scholars recently put it, "The fundamental problem in American higher education is no longer the *availability of content*, but rather the *availability of motivation*."[25] Motivation is a commodity colleges sell along with content and skills. We can't, ultimately, separate these three products. The more teacher excitement is there in lecture, the more students are likely to develop historical knowledge and skills. "Motivation depends on definable, concrete conditions, including physical co-presence."[26] As long as face-to-face expert instruction leads to intrinsic motivation and a pattern of lifelong learning, it will have an important place in history education.

Teacher enthusiasm also helps explain why mastery of historical content knowledge correlates so highly with great history teaching.[27] Numerous studies have shown that the more history a teacher knows, the better a history teacher they will be.[28] This seems a strange finding until we realize that mastery of history and enthusiasm for history both correlate with individual interest in history. A history teacher who loves her subject will be enthusiastic about teaching it *and* motivated to know a lot about it. It's not that knowing more history *leads* her to be a better teacher of history; it's that her passion for history leads to both more knowledge *and* more enthusiasm in talking about it.

The other major benefit of lecturing in history is that it is perhaps the best way to help students achieve "level 2" expertise. To understand what this means, let's take a moment to understand the three interrelated levels of history learning.

At level 1, history students *memorize*. They acquire knowledge of the atomic facts of history—names, dates, events—and embed them into their minds. They encode isolated bits of knowledge into their long-term memory for later retrieval into their working memory.

At level 2, history students *organize*. They connect the atomic facts into an ordered, coherent whole—often called a "knowledge structure"—a complex, deep understanding of a historical topic. At level 2, students unify historical knowledge into meaningful patterns and frameworks such that they can "find their way around" the discipline.[29] The blizzard of isolated historical facts from level 1 becomes a meaningful pattern of integrated knowledge at level 2.

At level 3, history students *utilize*. They apply their historical knowledge by developing new historical theories and interpretations. They put their knowledge to work. This is the ultimate goal of historical learning, but is impossible to achieve without the lower two levels. For instance, to develop an explanatory theory about the American Revolution (level 3), a historian would need a comprehensive knowledge of Revolutionary American history (level 2), and to have this knowledge the historian would need to know atomic facts (level 1), such as the meaning of terms like "Stamp Tax," "George Washington," and "Battle of Trenton."

Levels 2 and 1 are complementary in that a knowledge structure increases our capacity to memorize, encode, and retrieve facts within that structure. Indeed, we often learn facts simultaneously with learning frameworks—as when reading a historical monograph. There is a virtuous cycle to memorization: the more we know about a topic, the more we *can* know.[30] An expert on a topic has webs of connections, organizations, and groupings that aid, rather than impede, their memorization.[31]

Seeing these three levels as complementary helps us reconcile the seemingly opposed theories of John Dewey and E. D. Hirsch and end the fruitless debate between "progressive" and "traditional" approaches to history education. In the early twentieth century, education theorist Dewey argued that schools should move their focus away from rote memorization (level 1) and toward skills of instrumental value (level 3).[32] For decades, his acolytes, advocates of "progressive" education, have denigrated the value of facts in the name of educating for skills.[33] They pit level 3 against level 1, arguing that "facts don't matter," especially in our information age when we can quickly look those facts up online. Memorizing and learning material, they say, is unimportant since skills, not knowledge, are the ultimate goal.

This is an erroneous and dangerous way of thinking because you can't apply what you don't know. Imagine a surgeon slicing open a patient and then saying, "Everybody wait while I look up the location of the next artery." This would clearly be problematic for two reasons. First, the surgeon needs to have the fact in her head, not on her phone, to perform an operation and second, it's not the fact that matters, but the knowledge structure of the circulatory system—this can't be "looked up." It must be acquired through long, patient study and memorization of anatomical facts and their relations. The deep knowledge required for medical procedures is constructed out of the "raw material" of atomic facts and these facts must be in the surgeon's mind. They help the patient not at all when floating in cyberspace.

The same is true for history or any other field. My students can't develop historical theories without a deep knowledge of the era they are interpreting and this requires memorizing historical facts (and, yes, that means knowing names, dates, and events). There is a difference between knowledge and information and the information on one's phone can't be used to analyze the past. It must be knowledge in their heads. The same way we can't use a vocabulary word we don't know, we can't use a historical fact we don't know. Having a historical fact or vocabulary word on our phone is, literally, useless: "It is not sufficient only to attempt to teach general problem solving and thinking skills; the ability to think and solve problems requires a well-organized knowledge that is accessible in appropriate contexts."[34] We must acquire expertise before application is even possible. "The idea that expertise in a discipline is a general set of problem-solving skills that lack a content knowledge base to support them . . . is erroneous."[35] Dewey failed to realize that you can't apply what you don't know and you can't get to level 3 without first passing through levels 1 and 2.[36]

Unfortunately, too many historians have bought into Dewey's fallacy. Alia Wong reported in the October 21, 2015 issue of the *Atlantic* that a consensus was emerging among history educators that "instead of promoting the rote memorization of information . . . teachers should use a variety of primary-source materials and other writings, encouraging kids to analyze how these narratives are written."[37] These professors, while well intentioned in their desire to help students develop skills, fail to realize that it is impossible to engage in this application (level 3 interpretation of a document) without any understanding of the document's historical context (level 2 content understanding). Their suggestion that we use more documentary analysis in history teaching is something I agree with, but they err in thinking we can analyze a document without having underlying historical knowledge.

How could a student give a reasonable, insightful analysis of, say, a letter from a Civil War soldier to his mother without an understanding of secession, union, southern ideology, the overall progression of the war, abolitionism, and slavery? If we stop teaching facts, we kill the student's ability to effectively analyze documents. There is no level 3 without level 2, and there is no level 2 without level 1. We can't read documents without context and we can't have context without content, but misguided educators continue to try to convince us otherwise.

Many of Dewey's opponents have been guilty of the same basic error. Like Dewey, they make the mistake of pitting the three levels of learning against one another, rather than seeing them as complementary. For instance, E. D. Hirsch has been arguing for decades that the Deweyeite focus on "educating for skills" has been a disaster and accounts for the low performance of American students relative to those of other developed nations.[38] Schools need to get back to teaching content (levels 1 and 2) instead of teaching for application (level 3), says Hirsch. But like his nemesis, Hirsch fails to realize that teaching for content is inexorably bound up with teaching for skills. He is right that content knowledge is crucial, but fails to see how central the application of knowledge is to ultimate education goals.

More recently, some advocates of "deep learning" have set level 2 against level 1, arguing that we should be teaching students the "big picture" instead of "just facts."[39] Bestselling historian Jim Loewen complains that history teachers are so focused on minutiae that they are missing the forest for the trees; in fact, they are even missing the trees for the "twigs."[40] Loewen fails to realize that forests are made up of trees and trees are made up of twigs. To think we can understand a forest without trees (and twigs—atomic facts) is a fallacy. We can't create deep learning out of nothing, but only out of existing knowledge.[41] Memorization is necessary, but not sufficient, for deep learning.[42] "Mere memorization of facts" is not higher order thinking, but higher order thinking depends on it."[43] We can't sculpt without clay and we can't have deep knowledge without facts.

We learn, ultimately, at all three levels and should not pursue one to the exclusion of the other two. Before jettisoning "skills" or "facts" or "knowledge" in favor of one of the other two, we need to recognize their inherent interdependence. There is no application without knowledge and no knowledge without facts.

What does this have to do with lecture? It turns out that expert instruction is perhaps the best way to provide students with level 2 historical knowledge.[44] When historians teach for knowledge structures they take disjointed historical facts that students might have gotten in preparation for class or picked up haphazardly, and help tie them together into a unified framework. Students can get basic historical facts from textbooks or independent research, but they have a much harder time turning these isolated facts into coherent, meaningful patterns of knowledge. Expert instruction from an articulate, passionate master of history helps students develop complex understandings of historical topics such that their facts become unified. A gifted history teacher can create strong knowledge structures for students in ways that other activities cannot.[45] This means that lecture can provide the crucial link between mere historical facts (level 1) and historical skills (level 3). A motto for history teachers should be: *instruct for structure.*

Not surprisingly, instructing for structure works better when done in person. The more complex the information, the more necessary the face-to-face communication.[46] Since level 2 learning is the most complex, delicate, and difficult kind, it requires the most nuance, help, and interaction to acquire.

Since interactive communication is superior to other kinds of learning when we are dealing with complicated ideas, then we should spend our lecture time helping students acquire that complex, level 2 knowledge, particularly since in-person communication allows the interchange and questioning that will allow students to fill the gaps in their knowledge.[47]

There is an assumption among many educators, especially among advocates of the "flipped" classroom, that we should spend class time problem solving rather than learning.[48] The flipped classroom is a nice ideal, but less applicable to history.[49] Without coherent and robust knowledge structures, students won't be able to apply their knowledge in an effective, meaningful way. The flipped classroom is also not as novel as its advocates claim. For generations, students have been learning material in preparation for class (e.g., reading textbooks) and then spending class time completing homework assignments with peers and instructors (this is how virtually all of my K-12 math classes were conducted). Is the flipped classroom really such a game changer when it has already been employed for generations with such mixed results?

This is not to say that all history educators should always lecture. All learning methods have conditions under which they are effective (although direct instruction has fewer, explaining why it is often the classroom default), and for lecture to be effective, two conditions must be met.

First, the instructor must good at it. If a history teacher is, indeed, a drone with low enthusiasm and poor communication skills, then, they should seek out and employ alternative learning activities. The best instructors should keep instructing, but those who lack the ability should at the very least reduce their amount of expert instruction.

Second, lecture must be "broken." That is, to be maximally effective, instructors should break up their lectures with opportunities for students to generate, elaborate, reflect, and retrieve. This keeps students engaged, helps them retain their knowledge, and makes their learning experience meaningful and active.[50] Most attacks on lecture are actually attacks on *unbroken* lecture, but don't apply to the kind of lecturing that the best history teachers use.[51]

Breaking a historical lecture means regularly stopping to pose questions that require students to "generate" and "elaborate" (or "retrieve") answers. Learning in the service of answering a question has been found to be highly effective for increasing student curiosity, purpose, attention, and memory.[52]

Generation means letting students come up with their own independent answers to questions. This is particularly useful when done in a "just in time" (JIT) manner that has students try to answer a question before being given the answer by the professor. Generating an answer that has not been given, even when wrong, makes the student more open, curious, eager, and receptive to the correct answers.[53]

This JIT approach to history teaching gives us a practical way to synthesize the seemingly contradictory theories of Dewey and Hirsch. No, we can't engage in fruitful application until we have knowledge (Hirsch), but briefly struggling with problems before receiving solutions (Dewey) can help us be more receptive to the knowledge and solutions given later.[54] Active learning is a state of mind, not a behavior, and expert instruction in history becomes more active when students grapple with historical questions before being given answers.

Numerous studies have validated the value of the JIT teaching principle.[55] Priming students for learning through generation greatly enhances the effectiveness of expert instruction. Lecture is a highly effective learning tool so long as students have been prepared ahead of time by being forced to think about the material on their own.[56] Generation helps "plow the ground" of the student's mind for receiving the seeds of knowledge in lecture.[57] The JIT principle also helps students transition from being efficient to adaptive learners who can apply knowledge in a wide range of settings.[58] Priming students with JIT methods in this way is a simple, but extremely effective way to make historical learning stick and to improve student abilities to develop historical thinking skills.[59]

How would this work in practice? A history professor could begin the day's instruction in the 1930s United States by requiring students (individually or in groups) to come up with an explanation for the Great Depression. After holding students accountable for such generation (perhaps through cold calls), the professor could then instruct the class on the prevailing theories historians have come up with (e.g., monetarist or Keynesian explanations). Once students had thus "learned by example," the professor could then conduct a discussion with the class in which they had to determine which explanation of the Great Depression was most convincing and why. Priming (through JIT generation), learning, and applying would thus be interwoven for maximum effect.

While history instructors should have students engage in *generation* before instruction, they should have students engage in *elaboration* after instruction.[60] Elaboration means having students restate what they have learned in their own words. This can be as simple as requiring students to answer brief, summary questions over what they were just taught, giving them a review quiz over the material, having them explain the material to a partner (pair and share), calling on individual students to recapitulate, or even asking the class as a whole to shout back a short answer to a summary question. This forced practice retrieval cements learning in the same way that generation prepares them for it. One study found that such simple elaboration exercises could double student learning.[61]

Elaboration is essentially a way to interrupt forgetting. Our brains naturally lose most of what comes through working memory since it assumes that information in the working memory is unimportant unless we tell it otherwise. Retrieval is how we tell our brains the information is important and thereby arrest the forgetting process.[62] Post-learning elaboration is a great form of retrieval. (We should also encourage students to elaborate when reading by making short summaries of each paragraph or page in a notebook or in the margins.)

Teacher monitoring of elaboration can also be a powerful feedback tool that can enhance learning.[63] "Scholars of teaching have long known that the more rapid the feedback, the better; and there's nothing more immediate than feedback from a live audience. This is because interactions—the more immediate the better—engage people emotionally, and motivate them to act and to change."[64] A huge advantage of in-person lecture is that it allows for immediate clarification when students are confused. Getting "lost" in a lecture (because of the inability to create the level 2 knowledge structure that organizes and gives clarity to a topic) derails the whole learning process. It's far more common to get lost online when there is no possibility for getting clarification from an instructor. Questions and engagement also make the students feel needed, and a feeling of being needed leads to greater active engagement, pro-activity, and investment in one's education.[65]

The feedback also helps instructors improve their teaching. Professors can adjust and alter their instruction based on student cues and responses in ways that they can't when teaching to an online audience. College works best, "when it provides a thick environment of constant feedback."[66] In co-spatial environments, there is far more shared focus, less distraction, and more instructor control over the learning. The shared attention necessary for deep learning can be monitored and promoted by in-person instructors in ways that can't be done when students are watching a video. Online lecture is more likely to have students distracted by technology, entertainment, and other people. I've talked to students who say that they actually binge-watch Netflix shows while also watching their online class lectures—needless to say, these students are learning far less than their in-person counterparts. Multitasking *does* hurt learning and it's a danger for online courses.[67] By contrast, in-person classes are intrinsically less likely to have distractions since there is personal monitoring, interaction, and the natural tendency to focus when interacting with others face-to-face.

Combining these findings leads us to a pattern of education in which we surround segments of a broken lecture with opportunities for generation (before) and elaboration/retrieval (after).[68] This means that an ideal lecture class might go as follows:

- generation on history topic 1 →
- instruction (lecture) on history topic 1 →
- elaboration on history topic 1 →
- generation on history topic 2 →

- instruction on history topic 2 →
- elaboration on history topic 2 →
- generation on history topic 3 . . . and so on.

Thus, lecture is an important part of a history classroom experience, but is broken up by the crucial exercises of generation and elaboration. This broken, interactive approach to history lecturing has been proven to be highly effective in producing student learning. This pattern can turn good lecturers into great lecturers.

I conclude this chapter on a note of optimism. Teaching history using the lecture method has been under attack for generations, and yet the effectiveness of the method itself is its own response. Unquestionably and appropriately, face-to-face lecturing will continue to have an important place in the formal education of history students. The question is not whether lecture "works" (it does, as we saw above, under the right conditions), but whether or not we will make our lectures more effective by updating our approach using the findings of cognitive psychology. I hope that the above has not only justified lecture as an effective method for teaching history, but also given insight to teachers about how to make their lectures better.

NOTES

1. L. H. Summers, "What You (Really) Need to Know." *New York Times*, January 20, 2012. Available at http://www.nytimes.com/2012/01/22/education/edlife/the-21st-century-education.html; C. Evans, S. M. Kosslyn, & S. Kozhevnikov, "Cognitive Style as Environmentally Sensitive Individual Differences in Cognition: A Modern Synthesis and Applications in Education, Business, and Management." *Psychological Science in the Public Interest* 15/1 (May 2014): 3–33; G. Wood, "The Future of College? A Brash Tech Entrepreneur Thinks He Can Reinvent Higher Education by Stripping it Down to its Essence, Eliminating Lectures and Tenure Along with Football Games, Ivy-Covered Buildings, and Research Libraries. What If He''s Right?" *The Atlantic*, September 2014. Available at, https://www.theatlantic.com/magazine/archive/2014/09/the-future-of-college/375071/.

2. C. Wieman, *Improving How Universities Teach Science: Lessons from the Science Education Initiative* (Cambridge: Harvard University Press, 2017); S. Freeman, S. L. Eddy, M. McDonough, M. K. Smith, N. Okoroafor, H. Jordt, & M. P. Wenderoth, "Active Learning Increases Student Performance in Science, Engineering, and Mathematics." *Proceedings of the National Academy of Sciences of the United States* 111/23 (June 10, 2014): 8410–15.

3. Ironically, most of the anti-lecture diatribes I have heard have come from someone who was at that moment lecturing (undercutting what they say in the very act of saying it).

4. Wieman, *Improving How Universities Teach*; Summers, "What You (Really) Need to Know"; Wood, "The Future of College?"

5. J. Bransford, A. L. Brown, & R. R. Cocking, eds., *How People Learn: Brain, Mind, Experience, and School* (Washington D.C.: National Academies Press, 2000).

6. R. E. Mayer, "Should There Be a Three-Strikes Rule Against Pure Discovery Learning? The Case for Guided Methods of Instruction." *American Psychologist* 59 (2004): 15–17.

7. J. Hansen, "Seeking Principles, Not Fads, in Education." BYU-Idaho University Forum, January 15, 2017.

8. M. D. Miller, *Minds Online: Teaching Effectively with Technology* (Cambridge: Harvard University Press, 2014).

9. D. F. Chambliss & C. G. Takacs, *How College Works* (Cambridge: Harvard University Press, 2014); R. J. Marzano, D. Pickering, & J. E. Pollock, *Classroom Instruction that Works: Research-based Strategies for Increasing Student Achievement* (Alexandria, VA: Association for Supervision and Curriculum Development, 2001).

10. Miller, *Minds Online.*

11. Chambliss & Takacs, *How College Works.*

12. K. A. Ericsson, R. Krampe, & C. Tesch-Romer, "The Role of Deliberate Practice in the Acquisition of Expert Performance." *Psychological Review* 100 (1993): 363–406.

13. D. H. Pink, *Drive: The Surprising Truth About What Motivates Us* (New York: Penguin, 2011).

14. D. L. Schwartz, J. M. Tsang, & K. P. Blair, *The ABC's of How We Learn: 26 Scientifically Proven Approaches, How They Work, and When to Use Them* (New York: Norton, 2016).

15. Miller, *Minds Online.*

16. Chambliss & Takacs, *How College Works.*

17. Miller, *Minds Online.*

18. Schwartz et al., *ABC's of How We Learn.*

19. E. Loeb Bettinger, "Promises and Pitfalls of Online Education." *Brookings Institution Evidence Speaks Reports* 2, #15, June 9, 2017.

20. Miller, *Minds Online.*

21. Chambliss & Takacs, *How College Works*, 131–2.

22. Miller, *Minds Online.*

23. Schwartz et al., *ABC's of How We Learn.*

24. Chambliss and Takacs, *How College Works.*

25. Ibid., 173.

26. Ibid., 82.

27. Ken Bain, *What the Best College Teachers Do* (Cambridge: Harvard University Press, 2004).

28. E. D. Hirsch, *Why Knowledge Matters* (Cambridge: Harvard Education Press, 2016); D. L. Ball, M. H. Thames, & G. Phelps, "Content Knowledge for Teaching: What Makes It Special?" *Journal of Teacher Education* 59 (2008): 389.

29. Bransford et al., *How People Learn.*

30. Schwartz et al., *ABC's of How We Learn.*
31. Miller, *Minds Online.*
32. J. Dewey, *The School and Society: Being Three Lectures* (Chicago: University of Chicago Press, 1907); and *Democracy and Education: An Introduction to the Philosophy of Education* (New York: Macmillan, 1922).
33. Summers, "What You Really Need to Know."
34. Bransford et al., *How People Learn*, 153.
35. Ibid., 188.
36. Schwartz et al., *ABC's of How We Learn*; Mayer, "Should There Be a Three-Strikes Rule Against Pure Discovery Learning."
37. A. Wong, "History Class and the Fictions About Race in America." *The Atlantic*, October 21, 2015. Available at https://www.theatlantic.com/education/archive/2015/10/the-history-class-dilemma/411601/.
38. Hirsch, *Why Knowledge Matters*; E. D. Hirsch, *Cultural Literacy: What Every American Needs to Know* (New York: Vintage, 1987).
39. Summers, "What You Really Need to Know."
40. Wong, "History Class and the Fictions About Race in America."
41. Bransford et al., *How People Learn.*
42. P. C. Brown, *Make it Stick: The Science of Successful Learning* (Cambridge: Harvard University Press, 2014).
43. Miller, *Minds Online*, 89.
44. M. Storper & A. J. Venables, "Buzz: Face to Face Contact and the Urban Economy." *Journal of Economic Geography* 4 (2004): 351–70.
45. Bransford et al., *How People Learn.*
46. Storper and Venables, "Buzz," 367–8.
47. J. Gaspar & E. Glaeser, "Information Technology and the Future of Cities." *Journal of Urban Economics* 43/1 (1998): 136–56; and Mayer, "Should There Be a Three-Strikes Rule Against Pure Discovery Learning?"
48. Summers, "What You Really Need to Know."
49. Wieman, *Improving How Universities Teach Science*; and Freeman et al., "Active Learning Increases Student Performance in Science, Engineering, and Mathematics."
50. D. L. Schwartz & J. D. Bransford, "A Time for Telling." *Cognition and Instruction* 16 (1998): 475–522.
51. Wieman, *Improving How Universities Teach Science.*
52. Schwartz et al., *ABC's of How We Learn.*
53. Ibid.
54. Ibid.
55. Ibid.; Brown, *Make It Stick*; Bransford et al., *How People Learn*; G. M. Walton, G. L. Cohen, D. Cwir, & S. J. Spencer, "Mere belonging: The Power of Social Connections." *Journal of Personality and Social Psychology* 102 (2012): 513–32.
56. Schwartz and Bransford, "A Time for Telling."
57. Brown, *Make It Stick.*
58. Schwartz et al., *ABC's of How We Learn.*
59. Ibid.

60. Schwartz and Bransford, "A Time for Telling."

61. K. A. Ericsson, W. G. Chase, & S. Faloon, "Acquisition of a Memory Skill." *Science* 208 (1980): 1181–82.

62. Brown, *Make it Stick.*

63. J. Hattie & H. Timperley, "The Power of Feedback." *Review of Educational Research* 77 (2007): 81–112.

64. Chambliss and Takacs, *How College Works*, 131–2.

65. C. Christensen, "What I've Learned About Education," Faculty Luncheon at BYU-Idaho, February 16, 2017.

66. Chambliss and Takacs, *How College Works*, 132.

67. Miller, *Minds Online.*

68. Schwartz and Bransford, "A Time for Telling."

WORKS CITED

Ambrose, S. A., M. W. Bridges, M. DiPietro, M. C. Lovett, & M. K. Norman. 2010. *How Learning Works: Seven Research-Based Principles for Smart Teaching*. San Francisco: Jossey-Bass.

Asch, S. E. 1951. "Effects of Group Pressure upon the Modification and Distortion of Judgment." In *Groups, Leadership and Men*, H. Guetzkow ed. Pittsburgh: Carnegie Press: 177–90.

Bain, Ken. 2004. *What the Best College Teachers Do*. Cambridge: Harvard University Press.

Baldwin, C. & K. Clark. 2000. *Design Rules: The Power of Modularity*. Cambridge: MIT Press.

Ball, D. L., M. H. Thames, & G. Phelps. 2008. "Content Knowledge for Teaching: What Makes It Special?" *Journal of Teacher Education* 59: 389–407

Barron, B. 2003. "When Smart Groups Fail." *Journal of the Learning Sciences* 12: 307–59.

Battiston, D., J. B. Vidal, & T. Kirchmaier. 2017. "Is Distance Dead? Face-to-Face Communication and Productivity in Teams." Center for Economic Performance, London School of Economics, CEP Discussion Paper No 1473, March.

Bettinger, E., & S. Loeb. 2017. "Promises and Pitfalls of Online Education." *Brookings Institution Evidence Speaks Reports* 2, #15, June 9.

Bransford, J., A. L. Brown, & R. R. Cocking, eds. 2000. *How People Learn: Brain, Mind, Experience, and School*. Washington D.C.: National Academies Press.

Brown, P. C. 2014. *Make it Stick: The Science of Successful Learning*. Cambridge: Harvard University Press.

Cain, S. 2013. *Quiet: The Power of Introverts in a World that Can't Stop Talking*. New York: Random House.

Cairncross, F. 2001. *The Death of Distance 2.0: How the Communications Revolution Will Change our Lives*. Cambridge: Harvard Business School Press.

Campbell, E. M., H. Liao, A. Chuang, J. Zhou, & Y. Dong. 2017. "Hot Shots and Cool Reception? An Expanded View of Social Consequences for High Performers." *Journal of Applied Psychology* 102/5: 845–66.

Caplan, B. 2017. *The Case Against Education*. Princeton: Princeton University Press.
Chambliss, D. F. & C. G. Takacs. 2014. *How College Works*. Cambridge: Harvard University Press.
Christensen, C. 2017. "What I've Learned About Education." Faculty Luncheon at BYU-Idaho, February 16.
Christensen, C. & H. J. Eyring. 2011. *The Innovative University*. San Francisco: Wiley.
Christensen, C., M. B. Horn, & C. W. Johnson. 2011. *Disrupting Class*. New York: McGraw-Hill.
Cross, R., R. Rebele, & A. Grant. 2016. "Collaborative Overload." *Harvard Business Review*, January-February. Available at https://hbr.org/2016/01/collaborative-overload.
Crouch, C. H. & E. Mazur. 2001. "Peer instruction: Ten Years of Experience and Results." *American Journal of Physics* 69/9: 970–77.
Dewey, J. 1907. *The School and Society: Being Three Lectures*. Chicago: University of Chicago Press.
———. 1922. *Democracy and Education: An Introduction to the Philosophy of Education*. New York: Macmillan.
Duckworth, A. 2016. *Grit: The Power of Passion and Perseverance*. New York: Scribner.
Dweck, C. 2007. *Mindset: The New Psychology of Success*. New York: Ballantine.
Ericsson, K. A., R. Krampe, & C. Tesch-Romer. 1993. "The Role of Deliberate Practice in the Acquisition of Expert Performance." *Psychological Review* 100: 363–406.
Ericsson, K. A., W. G. Chase, & S. Faloon. 1980. "Acquisition of a Memory Skill." *Science* 208: 1181–82.
Evans, C., S. M. Kosslyn, S. Kozhevnikov. 2014. "Cognitive Style as Environmentally Sensitive Individual Differences in Cognition: A Modern Synthesis and Applications in Education, Business, and Management." *Psychological Science in the Public Interest* 15/1: 3–33.
Freeman, S., S. L. Eddy, M. McDonough, M. K. Smith, N. Okoroafor, H. Jordt, & M. P. Wenderoth. 2014. "Active Learning Increases Student Performance in Science, Engineering, and Mathematics." *Proceedings of the National Academy of Sciences of the United States* 111/23: 8410–15.
Gaspar, J. & E. Glaeser. 1998. "Information Technology and the Future of Cities." *Journal of Urban Economics* 43/1: 136–56.
Glaeser, E. 2011. *Triumph of the City: How Our Greatest Invention Makes Us Richer, Smarter, Greener, Healthier, and Happier*. New York: Penguin.
Grodal, S., A. J. Nelson, & R. M. Siino. 2015. "Help-Seeking and Help-Giving as an Organizational Routine: Continual Engagement in Innovative Work." *Academy of Management Journal* 58: 136–68.
Hansen, J. 2017. "Seeking Principles, not Fads, in Education." BYU-Idaho University Forum, January 15.
Hattie, J. & H. Timperley. 2007. "The Power of Feedback." *Review of Educational Research* 77: 81–112.
Hayek, F. A. 1945. "The Use of Knowledge in Society." *American Economic Review* 35/4: 519–30.

Hirsch, E. D. 1987. *Cultural Literacy: What Every American Needs to Know*. New York: Vintage.

———. 2016. *Why Knowledge Matters*. Cambridge: Harvard Education Press.

James, G. 2017. "'Collaboration' Creates Mediocrity, Not Excellence, According to Science." *Inc.com*, April 13. Available at: https://www.inc.com/geoffrey-james/collaboration-creates-mediocrity-not-excellence-according-to-science.html.

Johanes, P. & L. Lagerstrom. 2016. "Online Videos: What Every Instructor Should Know." *ASEE Annual Conference & Exposition*, New Orleans, LA, June 26.

Kirschner, P. A., J. Sweller, & R. E. Clark. 2006. "Why Minimal Guidance During Instruction Does Not Work: An Analysis of the Failure of Constructivist, Discovery, Problem-Based, Experiential, and Inquiry-Based Teaching." *Educational Psychologist* 41: 75–86.

Kizilceca, R. F. & G. L. Cohen. 2017. "Eight-Minute Self-Regulation Intervention Raises Educational Attainment at Scale in Individualist but not Collectivist Cultures." *Proceedings of the National Academy of Sciences,* February 27 (Early Edition): 1–6.

Knoke, K. 1996. *Bold New World: The Essential Road Map to the Twenty-First Century*. New York: Kodansha.

Labaree, D. F. 1997. "Public Goods, Private Goods: The American Struggle Over Educational Goals." *American Educational Research Journal* 54: 39–81.

Lagemann, E. C. 2000. *An Elusive Science: The Troubling History of Education Research*. Chicago: University of Chicago Press.

Lemann, N. 2016. "The Case for a New Kind of Core." *The Chronicle of Higher Education*, November 27.

Marzano, R. J., D. Pickering, & J. E. Pollock. 2001. *Classroom Instruction that Works: Research-based Strategies for Increasing Student Achievement*. Alexandria, VA: Association for Supervision and Curriculum Development.

Mayer, R. E. 2004. "Should There Be a Three-Strikes Rule Against Pure Discovery Learning? The Case for Guided Methods of Instruction." *American Psychologist* 59: 14–19.

Means, B., M. Bakia, & R. Murphy. 2014. *Learning Online: What Research Tells Us About Whether, When, and How*. New York: Routledge.

Milenko, M. 2017. "Strategizing Resources Leads to Improved Exam Scores, According to Stanford Scholars." *Stanford News*, May 8. Available at http://news.stanford.edu/2017/05/08/studying-strategically-equals-improved-exam-scores/.

Miller, M. D. 2014. *Minds Online: Teaching Effectively with Technology*. Cambridge: Harvard University Press.

Murray, C. & R. J. Herrnstein. 1996. *The Bell Curve: Intelligence and Class Structure in American Life*. New York: Free Press.

Negroponte, N. 1995. *Being Digital*. New York: Vintage.

Pink, D. H. 2011. *Drive: The Surprising Truth About What Motivates Us*. New York: Penguin.

Popper, K. R. 1973. *Objective Knowledge: An Evolutionary Approach*. New York: Oxford University Press.

———. 1957. *The Poverty of Historicism*. Boston: Beacon.

Schwartz, D. L. & J. D. Bransford. 1998. "A Time for Telling." *Cognition and Instruction* 16: 475–522.

Schwartz, D. L., C. C. Chase, M. A. Oppezzo, & D. B. Chin. 2011. "Practicing Versus Inventing with Contrasting Cases: The Effects of Telling First on Learning and Transfer." *Journal of Educational Psychology* 103: 759–75.

Schwartz, D. L., J. M. Tsang, & K. P. Blair. 2016. *The ABC's of How We Learn: 26 Scientifically Proven Approaches, How They Work, and When to Use Them.* New York: Norton.

Storper, M. & A. J. Venables. 2004. "Buzz: Face to Face Contact and the Urban Economy." *Journal of Economic Geography* 4: 351–70.

Summers, L. H. 2012. "What You (Really) Need to Know." *New York Times*, January 20. Available at http://www.nytimes.com/2012/01/22/education/edlife/the-21st-century-education.html.

Surowiecki, J. 2004. *The Wisdom of Crowds: Why the Many Are Smarter Than the Few and How Collective Wisdom Shapes Business, Economies, Societies and Nations.* New York: Doubleday.

Tetlock, P. 2006. *Expert Political Judgment.* Princeton: Princeton University Press.

Thille, C. 2016. "What the Science of Learning Indicates We Should Do Differently." In *Change We Must: Deciding the Future of Higher Education*, Matthew Goldstein and George Otte, eds. New York: Rosetta: 99–134.

Toffler, A. 1980. *The Third Wave.* New York: Bantam Books.

Walton, G. M., G. L. Cohen, D. Cwir, & S. J. Spencer. 2012. "Mere Belonging: The Power of Social Connections." *Journal of Personality and Social Psychology* 102: 513–32.

Wieman, C. 2017. *Improving How Universities Teach Science: Lessons from the Science Education Initiative.* Cambridge: Harvard University Press.

Wong, A. 2015. "History Class and the Fictions About Race in America." *The Atlantic*, October 21. Available at, https://www.theatlantic.com/education/archive/2015/10/the-history-class-dilemma/411601/.

Wood, G. 2014. "The Future of College? A Brash Tech Entrepreneur Thinks He Can Reinvent Higher Education by Stripping it Down to its Essence, Eliminating Lectures and Tenure Along with Football Games, Ivy-Covered Buildings, and Research Libraries. What If He's Right?" *The Atlantic*, September. Available at https://www.theatlantic.com/magazine/archive/2014/09/the-future-of-college/375071/.

Chapter 7

What Lectures Do Well

History Lectures as an Intermediate Pedagogy

Caroline R. Sherman

It is common to think about how lectures can fail, which is how we all know that lectures do not directly facilitate active learning. They are not effective, in themselves, at transferring detailed knowledge.[1] A captive audience may tend toward resentment, frustration, and annoyance, and lectures cannot cure their own limitations. A class that is going badly can almost never be set right by a lecture of any kind or quality. These facts may make the continued use of lectures in history courses appear suspect: Why do faculty continue to use a form of pedagogy with such obvious limitations? One could imagine that we do so because it is easy or satisfying for ourselves, rather than for the good of the students, and there is some justice to that charge.[2]

And yet academic norms of criticism have the tendency to see problems more clearly than benefits, as anyone who has tried to pass a proposal through a room full of faculty is aware. There are many things that lectures do well in the discipline of history, some of which are as easy to list as their most commonly cited deficiencies. Lectures are a good means to convey attitudes and to explain complex ideas clearly. They can provide context and model historical analysis in real time, as a mode of being. They are more under control and more predictable than other forms of teaching, which can be particularly good when dealing with sensitive or misunderstood subjects, but they can also adapt *in media res* to the audience. Lecturing also mimics one of the distinguishing poses of the historian: a person standing above the events, describing them with a view that was accessible to no one at the time, reifying abstractions. Through personification, the lecture can make that trick (and its limitations) more visible to the student than other pedagogical techniques. Finally, a lecturer can give students reasons to be interested in a topic that is strange to them by offering the most interesting details and assessments of a new topic. It is otherwise difficult for people to be interested in unfamiliar history—that is, the history of those people and events that do not seem immediately relevant to them.

As all of this suggests, lectures do serve some functions that are difficult to effectively replicate with other activities, especially at a time when students arrive at college with relatively low levels of historical knowledge, less experience with sustained reading, and greater interest in listening to podcasts.[3] Most significantly, lectures within the discipline of history are an *intermediate* pedagogy, well placed in the middle of the development of the discipline.[4] In fact, as a discipline we generally already structure history education such that intermediate learners are the most exposed to lectures.

After all, history begins in stories, and lectures can tell stories very well.[5] But the very young and those without any historical background cannot listen to stories for long without interrupting with necessary questions—questions that they cannot put aside because the lack of an answer will prevent them from following the narrative.[6] It is impossible for true beginners to read a lecture in the way that we want them to read a book—that is, at once engaging with the material presented while also maintaining an interior set of questions, objections, and thoughts that relate to it.[7] The best young or new students of history often need to exteriorize this dialogue as quickly as possible by raising hands to ask questions or make points.

The intermediate learner, however, should know enough history to be able to listen to a lecture without such pressing questions, and the function of the textbook in an introductory history course is (at least in theory) to ensure that all students have some rudimentary knowledge before the lecture begins.[8] These intermediate learners, taking advanced placement classes in high-school or entry-level classes in college, should have developed the skill of listening to a lecture while also sustaining a running internal commentary on it: evaluating the assertions, noting down the key information, and comparing what is discussed in the lecture to other information. Even if this is difficult at first, it becomes increasingly possible for them to cultivate a range of responses while listening to a lecture and to make the material their own. After all, taking notes on a lecture is not, in fact, a passive activity; the more engaged the student, the more active the experience of a lecture becomes. The supplementation of the lecture in the discussion section, in which themes of the lecture are sometimes challenged, helps to clarify that lectures should be responded to and not simply absorbed.

We use lectures as an intermediate pedagogy in history in part because they help to fill in a gap in the literature that historians produce. Professional historians do not write much that is aimed at the intermediate pedagogical development of students in the discipline. Monographs and articles are difficult to read without expertise. Textbooks and surveys do exist, but on any given specialized subject there are relatively few works to bridge the jump from intermediate to advanced work, particularly in topics such as legal history, institutional history, or uncelebrated periods of non-American national histories. The profession continues to expect the oral transfer of significant knowledge, whether in lecture, seminar discussions, or conversation, probably because there are so many potential topics of historical study, and so much depth to them; that it is more efficient to use oral methods in the classroom while preserving the writing hours of specialists to work at the edges of historical knowledge, rather than to create materials for intermediate learners. This is why the "mini-lecture" is almost impossible to eradicate, even from classrooms dedicated to active learning. Almost no one proposes that faculty should be entirely silent in the classroom.

Moreover, many students enroll in history courses in college for motives that the professional historian had to kill in him- or herself during graduate school: the search for identity, meaning, glory, or exemplary lessons. These themes will almost never feature in the writings of professional historians, no matter what level they aim at. And yet compelling lectures will often, by rhetorical necessity, touch at least a little on repetitive patterns or some of the emotional, suspenseful, tragic, or rueful aspects of the material, while also modeling epistemological uncertainty and neutrality, which means that the lecture can be a means to provide the sweet with the sour, the milk with the meat. As Jerome Bruner noted, stories are both vital for making sense of the world and also inevitably tangled in the moral order of things.[9] Lectures are an opportunity to simultaneously exploit and interrogate the moral assumptions of the stories we tell.

But then, as students near the end of their undergraduate degree, we tend to graduate them from the lecture, into seminars and senior theses, as a small sampling of the work that is expected in graduate school, where the experience of sitting through a lecture is relatively rare, confined, often, to special events or conferences, rather than being a mainstay of graduate education. Consequently, we acknowledge the intermediate learner cannot proceed to advanced studies by lectures alone: reading, writing, and discussing are vital for this, which is why many of the most advanced undergraduate courses and almost all graduate courses are seminars. After all, advanced students of history already know how many of the stories of the past turn out, which means that lectures will tend to be less compelling unless the narratives are particularly literary, obscure, or more analytical.[10]

Still, while history lectures are an intermediate pedagogy, they are also one of the highest arts of the discipline, a fact acknowledged by invited lecture series honoring the most distinguished historians. Although graduate students do sometimes lecture, best practices are usually to assign them the tasks of leading discussions and grading, with occasional guest lectures to learn the craft. After all, lecturing is memorable and entrancing when done well, insufferable when done badly, and so under ideal circumstances graduate students do not lead lecture courses until they have finished their training themselves.[11]

Although lectures have functioned as an intermediate pedagogy (and seminars as its advanced pedagogy) within history since its professionalization in the nineteenth century, it is possible that lectures are particularly useful at this moment in time. Not only have today's students demonstrated an interest in listening, through their embrace of podcasts as mentioned above, but lectures may respond to other contemporary national trends as well. Lectures can also be a kind of cognitive-behavior therapy: exposure to the catastrophic, and a means of looking at them without blinking. At a time when students report that they are struggling with anxiety and depression, increasingly interested in Stoicism, and finding resilience to be difficult, it may be profitable to model empathetic but sober reflections on the past.[12] The old saw about the "sage on the stage" is a good chastisement to our professorial pretensions, our own love of hearing ourselves talk, but it is also useful to ask how we think a sage would respond to the past we are describing or even to ask ourselves how we live with some of the things we know that humans have been capable of doing to each other. Lecturers can model acceptance of the past for students: how to integrate a forthright understanding of a nation's past into one's understanding of the world and oneself.

Indeed, we know that the study of history can nurture the ambitions of those who seek power over others; both Napoleon and Hitler were avid readers of history and conceived of themselves as students of its lessons.[13] But tendencies to mythologize, aggrandize, or abuse the past will not stop simply because professors abandon lectures; popular history will always be available to give their audiences what they want to hear, and our students consume this kind of history along with our courses. Some of this popular history claims to be secret histories that reveal conspiracies deliberately hidden by the academy. Lectures are at least an opportunity to provide an antidote to otherwise irrefutable grand conspiracies—rather than the petty conspiracies of which history does offer plenty of evidence—because a good lecturer can build a relationship of trust with the audience while also telling complicated, multicausal stories that are difficult to reinterpret in flat monocausal grand narratives.[14]

Still, it might seem imprudent to retain lectures at a time when history majors are declining because it is perceived that history majors are not employable.[15] Certainly all history courses should develop skills of historical analysis and writing. But too much rhetorical emphasis on skills implies that they matter more than the content and may inadvertently undermine history's already tenuous position. A lecture, at least, suggests that the content is important for its own sake. Lectures also designate the speaker as an expert in the field, which is not a bad thing to emphasize at a time when Americans are skeptical of expertise.[16] Perhaps it is not a coincidence that lectures have declined in pedagogical favor at the same time as the professoriate has started to diversify, it is not simply that new faculty may be more attuned to the ways that lectures can reenact structures of power, but it may also be that students may have difficulty listening to someone who they, at least subconsciously, do not believe could be an expert.[17] However, this is a difficulty that we should expect to subside over time.

Lecturing, like all teaching, depends on mutual goodwill and respect between audience and lecturer. In an age of polarization, generational divide, or hostility toward either academia or students, any given lecture may be less likely to work for everyone. But, as suggested earlier in the discussion about lectures and the historical pose, the weaknesses of lectures may also be strengths. All pedagogy is coercive and has "designs on" the student; it is impossible to create a curriculum that does not tell students what the instructor wants them to be, even if the instructor wants nothing more than for students to be themselves. But among pedagogical techniques, lectures are the most transparent and honest about their intentions, which is clarifying for the student. Moreover, ultimately lectures are unsatisfying, however enjoyable they may be for a while. Lecturers only have so much interesting material and may repeat jokes and stories from one class to another, and at some point we all tire of listening to someone else talk, no matter how compelling he or she is. The ultimate question for lecturers is simply this: When your students no longer want to listen to you, do they drop the discipline entirely, or do they find a way to talk back?

NOTES

1. They are not, on their own, particularly good means for the reception and retention of information unless supplemented with other exercises, as discussed in Hyrum Lewis's essay in this collection. James Lang's suggestions on how to improve retention are fully compatible with the use of lectures plus other techniques to improve learning. James Lang, *Small Teaching: Everyday Lessons from the Science of Learning* (San Francisco: Jossey-Bass, 2016).

2. Kevin Gannon has written some powerful and balanced criticisms of excessive use of lectures in history courses, most notably his "Lecture-Based Pedagogy and the Pitfalls of Expertise," Available at http://www.thetattooedprof.com/2018/01/16/lecture-based-pedagogy-and-the-pitfalls-of-expertise/.

3. U. S. Education Department, *The Nation's Report Card: U. S. History 2014*, does show slight improvement in eighth-grade knowledge of U.S. history. Available at https://www.nationsreportcard.gov/hgc_2014/#history. However, only 12 percent

of seniors were deemed proficient at U.S. history, with proficiency diminishing over time (20% of fourth graders, 18% of eighth graders, and only 12% of seniors). Edison Research found that young adults aged 18 to 34 were the most likely of all age groups to listen to a podcast in any given month. They also found that podcast consumption increased with education, "The Podcast Consumer 2017," Available at https://www.edisonresearch.com/wp-content/uploads/2017/04/Podcast-Consumer-2017.pdf.

4. See Hyrum Lewis's essay in this collection about the use of lectures for "Level 2" development, for example.

5. One avid defender of the lecture in history as a vehicle for stories is Honor Sachs, who contends that "lecturing is an art, a skill, a performance" that is (when done well) beloved by students but not by "the market research folks in higher ed tech." @drhonor, Twitter, February 22, 2017, Available at https://twitter.com/drhonor/status/834522734931668992 (accessed January 1, 2019). Christina Worth, "Before They Were Educators: Dr. Honor Sachs," *The Western Carolinian*, September 30, 2016. Available at http://thewesterncarolinian.com/?q=article/they-were-educators-dr-honor-sachs.

6. Teachers who cannot answer the many varied historical questions that young learners will ask may be tempted to focus on other subjects and skills that are more likely to be tested. Bruce VanSledright et al., "The End of History Education in Elementary Schools?" *Perspectives on History*, 2012. Available at https://www.historians.org/publications-and-directories/perspectives-on-history/may-2012/the-end-of-history-education-in-elementary-schools.

7. Jonathan Haidt suggests that "the human mind is a story processor, not a logic processor. Everyone loves a good story; every culture bathes its children in stories." Jonathan Haidt, *The Righteous Mind: Why Good People are Divided by Politics and Religion* (New York: Vintage, 2013), 287.

8. The fact that many students do not read the textbook before the lecture can be a significant problem in courses; if the professor uses the lecture to cover the material that students should have read, so that no one is lost, then the students will be less likely to read the textbook in the future, meaning that lectures increasingly try to recapitulate the groundwork of the textbook and can focus less on the intermediate work that the lecture is supposed to do.

9. Jerome Bruner, *Making Stories: Law, Literature, Life* (Cambridge: Harvard University Press, 2002); Paul C. Vitz, "The Use of Stories in Moral Development: New Psychological Reasons for an Old Education Method." *American Psychologist* 45/6 (1990): 709–20.

10. The tension between narrative surprise and student preparation is familiar to everyone who prepares a syllabus in history, since lectures are dull if they repeat material that students are supposed to have read; yet if students have not done the reading, then the lecture may be more difficult to follow.

11. Needless to say, in real circumstances, graduate students often adjunct as lecturers well before they finish their dissertations.

12. Jean Twenge et al., "Increases in Depressive Symptoms, Suicide-Related Outcomes, and Suicide Rates Among U.S. Adolescents After 2010 and Links to Increased New Media Screen Time," *Clinical Psychological Science* 6/1 (2017):

3–17. Donald Robertson, *The Philosophy of Cognitive-Behavioural Therapy (CBT): Stoic Philosophy as Rational and Cognitive Psychotherapy* (London: Karnac, 2010); Chiara Sulprizio, "Why is Stoicism Having a Cultural Moment?" 2015. Available at https://eidolon.pub/why-is-stoicism-having-a-cultural-moment-5f0e9963d560.

13. Napoleon's sense of history was largely self-taught, but Hitler claimed to have been deeply inspired by his high-school history teacher, the captivating Leopold Poetsch.

14. On the psychological functions of conspiracy histories, see Jan-Willem van Prooijen, *The Psychology of Conspiracy Theories* (Oxon: Routledge, 2018).

15. Paul B. Sturtevant, "History Is Not a Useless Major: Fighting Myths with Data," *Perspectives on History*, April 1, 2017. Available at https://www.historians.org/publications-and-directories/perspectives-on-history/april-2017/history-is-not-a-useless-major-fighting-myths-with-data; Benjamin Schmidt, "The Humanities Are in Crisis." 2019. Available at https://www.theatlantic.com/ideas/archive/2018/08/the-humanities-face-a-crisisof-confidence/567565.

16. Tom Nichols, *The Death of Expertise: The Campaign Against Established Knowledge and Why It Matters* (New York: Oxford University Press, 2017); Susan Jacoby, *The Age of American Unreason in a Culture of Lies* (New York: Vintage, 2018); Richard Hofstadter, *Anti-Intellectualism in American Life* (New York: Vintage, 1966).

17. Tyler Stovall, "Diversity and Segregation: Progress and Challenges in the Struggle for an Inclusive Historical Community," *Perspectives on History*, March 10, 2017. Available at https://www.historians.org/publications-and-directories/perspectives-on-history/march-2017/diversity-and-segregation-progress-and-challenges-in-the-struggle-for-an-inclusive-historical-community. On the political effects of pedagogy, see Paulo Freire, *Pedagogy of the Oppressed: 50th Anniversary Edition* (London: Bloomsbury Academic, 2018).

WORKS CITED

Bruner, Jerome. 2002. *Making Stories: Law, Literature, Life.* Cambridge: Harvard University Press.

Edison Research. 2017. "The Podcast Consumer 2017." Available at https://www.edisonresearch.com/wp-content/uploads/2017/04/Podcast-Consumer-2017.pdf.

Freire, Paulo. 2018. *Pedagogy of the Oppressed: 50th Anniversary Edition.* London: Bloomsbury Academic.

Gannon, Kevin. 2019. "Lecture-Based Pedagogy and the Pitfalls of Expertise." Available at http://www.thetattooedprof.com/2018/01/16/lecture-based-pedagogy-and-the-pitfalls-of-expertise/.

Haidt, Jonathan. 2013. *The Righteous Mind: Why Good People are Divided by Politics and Religion.* New York: Vintage.

Hofstadter, Richard. 1966. *Anti-Intellectualism in American Life.* New York: Vintage.

Jacoby, Susan. 2018. *The Age of American Unreason in a Culture of Lies.* New York: Vintage.

Lang, James. 2016. *Small Teaching: Everyday Lessons from the Science of Learning.* San Francisco: Jossey-Bass.

Nichols, Tom. 2017. *The Death of Expertise: The Campaign Against Established Knowledge and Why It Matters.* New York: Oxford University Press.

Robertson, Donald. 2010. *The Philosophy of Cognitive-Behavioural Therapy (CBT): Stoic Philosophy as Rational and Cognitive Psychotherapy.* London: Karnac.

Sachs, Honor. 2019. Available at https://twitter.com/drhonor/status/834522734931668992.

Schmidt, Benjamin. 2019. "The Humanities Are in Crisis." Available at https://www.theatlantic.com/ideas/archive/2018/08/the-humanities-face-a-crisisof-confidence/567565.

Stovall, Tyler. 2017. "Diversity and Segregation: Progress and Challenges in the Struggle for an Inclusive Historical Community." *Perspectives on History*, March 10. Available at https://www.historians.org/publications-and-directories/perspectives-on-history/march-2017/diversity-and-segregation-progress-and-challenges-in-the-struggle-for-an-inclusive-historical-community.

Sturtevant, Paul B. 2017. "History Is Not a Useless Major: Fighting Myths with Data." *Perspectives on History*, April 1. Available at https://www.historians.org/publications-and-directories/perspectives-on-history/april-2017/history-is-not-a-useless-major-fighting-myths-with-data.

Sulprizio, Chiara. 2015. "Why is Stoicism Having a Cultural Moment?" Available at https://eidolon.pub/why-is-stoicism-having-a-cultural-moment-5f0e9963d560.

Twenge, Jean M., et al. 2017. "Increases in Depressive Symptoms, Suicide-Related Outcomes, and Suicide Rates Among U.S. Adolescents After 2010 and Links to Increased New Media Screen Time." *Clinical Psychological Science* 6/1: 3–17.

U. S. Education Department. *The Nation's Report Card: U. S. History 2014.* Available at https://www.nationsreportcard.gov/hgc_2014/#history.

Van Prooijen, Jan-Willem. 2018. *The Psychology of Conspiracy Theories.* Oxon: Routledge.

VanSledright, Bruce, et al. 2012. "The End of History Education in Elementary Schools?" *Perspectives on History.* Available at https://www.historians.org/publications-and-directories/perspectives-on-history/may-2012/the-end-of-history-education-in-elementary-schools.

Vitz, Paul C. 1990. "The Use of Stories in Moral Development: New Psychological Reasons for an Old Education Method." *American Psychologist* 45/6: 709–20.

Worth, Christina. 2016. "Before They Were Educators: Dr. Honor Sachs." *The Western Carolinian*, September 30. Available at http://thewesterncarolinian.com/?q=article/they-were-educators-dr-honor-sachs.

Chapter 8

Lecturing and Media Studies in the Twenty-First Century

Karen P. Burke

One morning, during my usual routine of walking to class, I overheard a conversation between two undergraduate students. One student enthusiastically said to the other student, "You don't have to go to *that* class, it's a *lecture* class!" The two students let out a boisterous laugh, high-fived one another, and went about their way. This encounter is not new; however, it speaks to the need to investigate the relevancy of lecturing in higher education, in particular, the field of media studies. As higher education continues to experience changes in the face of enrollment and, important new advances in media and technology, does lecturing have a place in media studies? Is the practice no longer in sync with ways in which twenty-first-century students access information, learn, think, and apply knowledge? Is the screen and the fragmented world of social media a better learning platform than the lecture hall? This chapter will address these questions and investigate the viability of lecturing in media studies. This chapter will be divided into five subsections: (1) the twenty-first-century postsecondary student; (2) twenty-first-century educational technology; (3) twenty-first-century media and technology; and (4) media studies and the "sage on the side."[1]

THE TWENTY-FIRST-CENTURY POSTSECONDARY STUDENT

A review of the literature, unequivocally, demonstrates that the U.S. college and university student is changing.[2] The National Center for Educational Statistics (NCES) projects that the number of students enrolling in colleges in the fall 2020 is estimated at 20 million, which is 6 million more students compared to 1990.[3] More students are attending college full time. Women have out-enrolled men at the undergraduate level since the 1970s and the 2020 projected data does not suggest that this trend is changing.[4] The majority of students enrolling at colleges remain 18 to 24 year-of age; however, a stagnation and decline in traditional-age students graduating from public and nonpublic high schools is reported.[5] More adults and veterans are enrolling at colleges.[6] In 2016, the NCES predicted that between 1990 and 2020, the 25 to 30+ year-old student cohort is expected to increase by 45.89 percent. This change is significant because adult students are often better prepared to handle the rigor of higher education and learn differently.[7]

Ethnic diversity is more representative on college campuses today. The NCES reports that the percentage of black, Hispanic, Asian/Pacific Islander, and American Indian/Alaska Native students is increasing while the percentage of white students is decreasing.[8] A 2017 Gallup Poll reports that more students who self-identify as lesbian, gay, bisexual, and transgendered are attending college. Campus Pride found that 267 colleges offer gender-inclusive housing (students can request a roommate of any gender). This type of inclusiveness of students' sexual orientation was nonexistent in 1990.

According to the Postsecondary National Policy Institute, first-generation college students are on the rise by 22 percent compared to 1992.[9] By the 2011–12 academic year, 34 percent of undergraduates were first-generation students. Scholars report that first-generation college students differ from other student because they: enter college with financial concerns; are likely to have food insecurity; work outside of school and are likely to send money home to support their families; have different motivations for attending college; are more likely to attend lower performing high schools, study alone, and experience less academic and emotional support from their families and communities; have feelings of anxiety and isolation; and, feel socially and emotionally marginalized.[10] These challenges create obvious hardships for first-generation students. Colleges have taken steps to ensure freshmen and first-generation student success by implementing first-year experience programs.[11] And, faculty are adapting their curriculum for first-generation students through a variety of practices.[12]

Students declaring disabilities and seeking learning accommodations is becoming more common. According to Belkin, "From 2010-11 to last year, the number of accommodations requests jumped 171%, while the number of people taking [SAT and PSAT among other tests] increased 22%. Last year, 94% of those requests were approved."[13] This change is noteworthy because both faculty and students may face new challenging. For example, faculty may need to accommodate students' need for disability-related absences and additional time to take test, complete assignments, and write essays; students with disabilities may have to adjust to new assistive technology; and both parties may have to coordinate time outside of class to cover difficult or missed materials.

The percentage of full-time undergraduates seeking financial aid grew to 85 percent during the 2015–2016 academic year.[14] Allen and Seam also claim that the number of students enrolled in online classes and distance learning is rapidly increasing.[15] The point here is this: if students have to decide between buying a book or food, students will choose food; if the lecture content is tied to books students do not have, or if students are underprepared to meet the challenges of college, this will impede the students' success. Furthermore, older and digital savvy students have a different skill sets and learning styles. For these reasons alone, academia has to look at new and inventive ways to meet the needs of twenty-first-century students.

TWENTY-FIRST-CENTURY EDUCATIONAL TECHNOLOGY

The last thirty years has experienced an explosion of new educational technology. In 1990, flipcharts and chalk boards were used in the classroom. Reams of paper were flying out of printers and into students' hands and final course grades were handwritten and delivered. Today, teaching occurs in chalk-and-paper-free settings. High-tech classroom bring a variety of educational materials to faculty and students at the push of a button. Courses are supported by virtual environments and course management systems, including Blackboard Learn and Canvas, which are expanding across campuses throughout the nation.[16]

With these new tools, educational resources are added for students to access online. Resources are comprised from a variety of formats including text, images, audio, animation, hyperlinks, slides, study tools, and discussion boards. Syllabi, course goals and objectives, and learning outcomes are available to refer back to whenever needed. Assignments, practice quizzes, and exams are posted, accessed by students, completed, and uploaded for evaluation. Students see their grades in real time. And, faculty virtually design and redesign their curriculum based on students' needs. Faculty and students are no longer confined to time, space, or location.

Facebook, Twitter, and blogs, to name a few new technologies, have been successfully introduced into curriculum design as well. Shaw reports that using Facebook in the classroom creates a sense of community and promotes collaboration; enhances communication between instructors and students; develops computer literacy and language skills; incorporates current student culture into the learning environment; and works well when sharing and discussing current events in the news.[17] Chawinga found that when designed effectively, students' use of Twitter and blogs in the classroom encourages: sharing and discussing course materials; posting course reflections; and interacting among students and professor 24/7.[18] History has shown technology and education has always had a symbiotic relationship; therefore, this ongoing alliance is not surprising.

TWENTY-FIRST-CENTURY MEDIA AND TECHNOLOGY

Media studies is an academic discipline that investigates various media and how it influences and shapes culture. The use of media and the U.S. federal policies governing the industry and changing technology informs and drives the discipline. The media is a meaning-making industry. It consists of a distinct group of experts who carefully and deliberately craft and distribute messages with intended meaning, aimed at specific audiences, in order to meet their own agendas. *All* information is derived from the media which makes them an extremely powerful institution.

The 1996 Telecommunication Act, signed by President Clinton, was the first major overall of the media industry since the Communications Act of 1934. This act was a game changer because it allowed media cross-ownership permitting businesses to own and operate multiple media entities. This act allowed the industry to be controlled by a handful of American oligopolies including Alphabet; Comcast; The Walt Disney Company; AT&T Entertainment Group; and News Corporation Ltd./21st-Century Fox.[19] Many scholars argue that media consolidation is unhealthy for democracy because it depends on the free and robust exchange of ideas; information that is presented in coherent and meaningful ways; access to important and useful information so that one can make well-information and intelligent decision; and, independent thought absent of commercial interests and personal agendas.[20]

It is important to note that the link between media oligopolies and technology is impregnable. For example, Alphabet is the world's fourth largest technology company with a market value of $766.3 billion and is the parent company of Google.[21] The remaining four companies are as equally anchored to technology, for-profit, and financially funded by advertising.

Technology refers to the tools and machines that are used to solve problems and achieve goals.[22] According to Bates, technology is used to support teaching, yet, technologies "do not of themselves communicate or create meaning. They just sit there until commanded to do something . . . or until a person starts to interact with the technology."[23]

Bates's distinction between media and technology is important. Technology operates at someone's will. In this case, technology functions as an aid in the production, distribution, and acceleration of messages. Its purpose is to be used as a tool of manipulation by the media. The media's function is to create meaning-intended messages for specific target audiences. The media's purpose is to make money by selling target audiences to advertisers. Media does not exist without the technology.

What has happened to media and technology in the last thirty years is remarkable. Consider for a moment how the World Wide Web, MP3 audio format, broadband, Bluetooth, hyperlinks, graphic processors, browsers, smartphones, tablets, and watches have changed and colonized our culture. We now have social media platforms that focus on the individual rather than the collective. Social media influences what we think about. We have new media being used to influence our behavior. New media beckons us; takes up mental space; alters our perceptions and opinions; reinforces our beliefs; collects information about us only to be used to sway us later; serves up the next best thing; shortens our attention spans; changes how we communicate, think, learn, and access information.[24] These very changes and their consequences are the focal point of media studies' teaching and scholarship.

MEDIA STUDIES AND THE "SAGE ON THE SIDE"

The Center for Innovation in Teaching and Learning (CITL) claims that lecturing is still the most widely used method of instruction across college campuses. However, the debate about lecturing as a viable pedagogy remains.[25] To begin, lecturing is considered a behaviorist teaching practice.

Those operating within a behaviorist paradigm believe that that purpose of education is to transfer knowledge, deliver instruction, and transmit information from the teacher to the learner. The behaviorist teaching/learning process, then, consists of direct instruction of facts that exist outside of the student, memorization of those facts, passive learning, and extrinsic reinforcement. The environment is teacher centered, and the role of the educator is to transmit information in the most efficient manner possible, often through lectures.[26]

McKeachie and Syinicki argue that lectures are good for presenting up-to-date information; summarizing material; adapting material to the background and interest of a particular group of students; focusing on concepts; and works best when it builds bridges between student's knowledge base and new material and subject matters.[27] Heward concurs that the lecture enables coverage of material not yet available in published form; can be used to supplement or elaborate course content; provides flexibility; can be personalized and customized to meet students' needs, interests and backgrounds; and can be motivating by sharing one's enthusiasm and commitment to the discipline.[28]

By contrast, active or student-centered curriculum design is considered a constructivist teaching practice. Teachers operating within a constructivist paradigm believe that the purpose of education is to help learners construct knowledge by building on individuals' schemas or cognitive structures to represent their meaning of the world.[29] Central to a constructivist teaching and learning approach is the belief that students play an active role in constructing their knowledge by linking new knowledge with past experiences.

Those who espouse a constructivist approach believe that active learning is more meaningful than passive learning—that is, learning which is delivered or transmitted to the learner. Learning is an active, constructive, and goal-oriented process that is dependent upon the mental activities of the learner.[30] The learning environment in a constructivist paradigm is student centered, and the role of the educator is to create a learning environment that builds upon learners' existing knowledge and experiential base, because their understanding and interpretation of specific content depends upon their prior knowledge and experience.[31]

The Berkeley Center for Teaching and Learning favors active learning paradigm because, they claim, the rigid lecture approach does not consider integrating additional pedagogical methodologies, does not customize learning for students, and will lead to student failure; yet, they do not dismiss lecturing altogether. They state:

> If we trouble the lecture debate by not defining it as an either/or (e.g., either I lecture, or use active learning), and instead as a both/and (e.g., what happens if I couple together a lecture with active learning?) something very different emerges—an understanding and appreciation for the complexity of employing lecture as a teaching tool. The essential component of its success is in recognizing that there is no one way to lecture, and not all lectures need to look alike in form, function, or duration. In fact, there is an entire body of literature on "Interactive Lecturing" that brings together aspects of active learning, formative assessment, and others to move lecture away from a monologue towards a dialogic experience where students actively engage in the lecture component of the course.[32]

With this said, the research shows that most educational institutions are shifting away from lecturing and moving toward more active learning models.[33] Studies show that today's college graduates have spent less than 5,000 hours of their lives reading text, while they have spent over 10,000 hours playing video games and 20,000 hours watching television.[34] This generation has been characterized as digital natives: a person that has grown up immersed in technology and consider digital technology as if it was a second language.

One cannot deny that a generational technological gap exists between veteran faculty and today's students. Effective educators look to their students for guidance and instruction in order to lessen the divide. This activity alone fosters connection and communication between faculty and students. As the above research suggests, modernizing pedagogy creates an effective and successful learning environment.

Media studies is a theory-driven discipline which requires critical reading, thinking, researching, writing, and presentation skills for students' success. As a result of the changing student, new developments in educational technology, and the enormous pool of media and technology resources to use in the classroom, media studies curriculum favors both lecturing and project-based curriculum design with an emphasis on the latter. As Zakrajsek suggests, an effective educator must not be the "sage on the stage" nor "the guide on the side" but rather, a "sage on the side."[35] He claims that faculty members must be the experts in the room otherwise they serve only as facilitators and vice versa. Furthermore, curriculum design works best when it adapts to student changes, challenges, needs, and technologically skills; hence, serving as both the expert through lecturing and the facilitator through project-based learning, serves the field of media studies well.[36]

Project-based curriculum design focuses on solving real-world problems which have local, national, and/or global significance. Once a problem is identified, students develop and demonstrate their understanding of the problem by gathering information from various sources, propose and design realistic solutions to solve that problem, and end with a problem/solution-based product or presentation.[37] The combination of lecturing and project-based learning work well in the field of media studies because theory acquisition is acquire through lecturing and critical reading (the sage on the stage), learning activities are faculty-guided (guide on the side) and, theory is applied through active, self-directed projects that have immediate world applications (student centered).[38] Through this pedagogy students quickly invest in their learning process; take great pride and ownership in the problem/solution-based work; learn that knowledge is power; understand their application of knowledge is important and relevant; and discover they can make positive change.

Lecturing coupled with project-based learning is the key to students' and faculty's success. Finding new ways to ensure and assess learning by designing curriculum based on timely, real-world problems, allows students to make informed decision about media in their own lives, recognize their ability to be change agents in a culture saturated and controlled by media, and contribute to media literacy, and the greater world in positive, productive, and meaningful ways.

NOTES

1. K-State, "The New Science of Learning: Concepts and Strategies for Changing the Way We Teach." 2014. Available at https://www.youtube.com/watch?v=P3IcDzr-tK4.

2. National Center for Education Statistics, "Total Fall Enrollment in Degree-Granting Postsecondary Institutions, by Attendance Status, Sex of Student, and

Control of Institution: Selected Years, 1947 through 2027." 2017. Available at https://nces.ed.gov/programs/digest/d17/tables/dt17_303.10.asp.

3. Ibid.

4. D. Borzelleca, "The Male-Female Ratio in College." *Forbes* (2012). Available at https://www.forbes.com/sites/ccap/2012/02/16/the-male-female-ratio-in-college/#18be4bf0fa52.

5. P. Bransberger, "Demographics, High School Graduates & Higher Education Demand." *Knocking at the College Door* (2018). Available at https://static1.squarespace.com/static/57f269e19de4bb8a69b470ae/t/5b57353ff950b7329f3c9856/1532441920461/July+24+2018+PBransberger+Slides+for+web.pdf.

6. Institute for Veterans and Military Families, "I am a Post-9/11 Student Veteran." *Student Veterans of America* (2017). Available at https://ivmf.syracuse.edu/wp-content/uploads/2017/06/I-AM-A-POST-911-Student-Veteran-REPORT.pdf.

7. A. G. Palis & P. A. Quiros, "Adult Learning Principles and Presentation Pearls." *Middle East Afr J Ophthalmol* 21/2 (2014): 114–22.

8. National Center for Education Statistics, "Total Fall Enrollment in Degree-Granting Postsecondary Institutions, by Attendance Status, Sex of Student, and Control of Institution: Selected Years, Selected Years, 1970 through 2026" (2016). Available at https://nces.ed.gov/programs/digest/d16/tables/dt16_303.40.asp?current=yes.

9. National Center for Education Statistics, "Status and Trends in the Education of Racial and Ethnic Groups 2016." (2016). Available at https://nces.ed.gov/pubs2016/2016007.pdf.

10. B. Galina, "Teaching First-Generation College Students." *Center for Teaching*. Available at https://cft.vanderbilt.edu/guides-sub-pages/teaching-first-generation-college-students/; A. C. Wilkins, "New in Class: First-Generation Students and the Unseen Costs of College." *Phi Kappa Phi Forum* (2018). Available at https://go.galegroup.com/ps/i.do?p=AONE&sw=w&u=googlescholar&v=2.1&it=r&id=GALE%7CA562377074&sid=classroomWidget&asid=632c45c1.

11. Southern Connecticut State University, "First Year Experience (FYE) Program." (2018). Available at https://www.southernct.edu/offices/first-year-experience/; C. Teniente-Matson, "Advice: 6 Steps to Help First Generation Students Succeed." *Higher Education* 65/26 (2018). Available at https://www-chronicle-com.scsu.idm.oclc.org/article/Advice-6-Steps-to-Help/242659.

12. See K-State for more information on faculty practices designed to ensure first-generation student success.

13. D. Belkin, "Colleges Bend the Rules for More Students, Give Them Extra Help." *The Wall Street Journal* (2018). Available at https://www.wsj.com/articles/colleges-bend-the-rules-for-more-students-give-them-extra-help-1527154200.

14. National Center for Education Statistics, *Financial Aid. Fast Facts.* (2018). Available at https://nces.ed.gov/fastfacts/display.asp?id=31.

15. E. Allen & J. Seaman, "Digital Learning Compass: Distance Education Enrollment Report 2017." *Babson Survey Group* (2017). Available at https://onlinelearningsurvey.com/reports/digtiallearningcompassenrollment2017.pdf.

16. S. Jaschik, *2018 Survey of Faculty Attitudes on Technology: A Study by Inside Higher Ed and Gallup* (2018). Available at https://www.insidehighered.com/system/files/media/IHE_2018_Survey_Faculty_Technology.pdf.

17. A. Shaw, "What is 21st Century Education? 21st Century Schools." (2009). Available at https:www.21stcenturyschools.com/What_is_21st_Century_Education.htm.

18. W. D. Chawinga, "Taking Social Media to a University Classroom: Teaching and Learning Using Twitter and Blogs." *International Journal of Educational Technology in Higher Education* 14/3 (2017). Available at https://educationaltechnologyjournal.springeropen.com/articles/10.1186/s41239-017-0041-6.

19. K. Stoller, "The World's Largest Tech Companies 2018: Apple, Samsung Take Top Spots Again." *Forbes*. Available at https://www.forbes.com/sites/kristinstoller/2018/06/06/worlds-largest-tech-companies-2018-global-2000/#28d5b4e14de6.

20. S. Jhally, "A Letter from Sut Jhally." *Media Education Foundation* (2011). Available at http://www.mediaed.org/letter-from-sut-jhally/; B. Moyers, "How Media Consolidation Threatens Democracy: 857 Channels (and Nothing on)." *Moyers* (2017). Available at https://billmoyers.com/story/media-consolidation-should-anyone-care/; R. W. McChesney, *Digital disconnect: How Capitalism is Turning the Internet Against Democracy* (New York: The New Press, 2013); J. Nichols & R. McChesney, *Our Media, Not Theirs: The Democratic Struggle Against Corporate Media* (New York: Seven Stories, 2002).

21. Stoller, "The World's Largest Tech Companies 2018."

22. University of Washington Center for Teaching and Learning, "Teaching with Technology." (2018). Available at http://www.washington.edu/teaching/teaching-resources/engaging-students-in-learning/teaching-with-technology-2/.

23. Computer Business Review, "What is Google." (2018). Available at https://www.cbronline.com/what-is/what-is-google/.

24. S. Briggs, "6 Ways Digital Media Impacts the Brain." *Open Colleges* (2016). Available at https://www.opencolleges.edu.au/informed/features/5-ways-digital-media-impacts-brain/; L. Belanger, "10 Ways Technology Hijacks our Behavior." *Entrepreneur* (2018). Available at https://www.entrepreneur.com/article/311284.

25. Berkeley Center for Teaching and Learning, *Remember and Understand. UC Berkeley: Division of Undergraduate Education* (2018). Available at https://teaching.berkeley.edu/resources/engage/remember-and-understand; T. J. Devlin, C. R. Feldhaus and K. M. Bentrem, "The Evolving Classroom: A Study of Traditional and Technology-Based Instruction in a STEM Classroom." *Journal of Technology Education* 25/1 (2013): 34–54; W. Heward, "Guided Notes–Improving the Effectiveness of Your Lectures." *Stanford University*. Available at https://tomprof.stanford.edu/posting/495.

26. W. McKeachie and M. Synicki, *McKeachie's Teaching Tips* (Boston: Cengage Learning, 2013).

27. Ibid.

28. Heward, "Guided Notes–Improving the Effectiveness of Your Lectures."

29. R. M. Cervero, *Effective Continuing Education for Professionals* (San Francisco: Jossey-Bass, 1988).

30. Ibid.

31. McKeachi and Synicki, *McKeachie's Teaching Tips*.

32. Ibid.

33. Berkeley Center for Teaching and Learning, *Remember and Understand*; Devlin, Feldhaus and Bentrem, "The Evolving Classroom."

34. M. Presnky, *Teaching Digital Natives* (Thousand Oaks: Corwin, 2010).

35. Todd Zakrajsek, "K-State Teaching and Learning Center." (2014). Available at https://www.youtube.com/watch?v=P3IcDzr-tK4.

36. Ibid.

37. J. Sepulveda, "Introduction: Project-Based Learning." *University of Oregon* (2016). Available at https://learn.canvas.net/courses/1123files/481829/download?verifier...wrap=1; Stanford University. 2018, "Project-Based Learning." *Stanford Teaching Common*. Available at https://teachingcommons.stanford.edu/resources/learning/learning-activities/project-based-learning.

38. Ibid.

WORKS CITED

Allen, E. & J. Seaman. 2017. "Digital Learning Compass: Distance Education Enrollment Report 2017." *Babson Survey Group*. Available at https://onlinelearningsurvey.com/reports/digtiallearningcompassenrollment2017.pdf.

Bates, T. n.d. *Teaching the Digital Age: Guidelines for Designing Teaching and Learning*. Available at https://opentextbc.ca/teachinginadigitalage/.

Belanger, L. 2018. "10 Ways Technology Hijacks our Behavior." *Entrepreneur*. Available at https://www.entrepreneur.com/article/311284.

Belkin, D. 2018. "Colleges Bend the Rules for More Students, Give Them Extra Help." *The Wall Street Journal*. Available at https://www.wsj.com/articles/colleges-bend-the-rules-for-more-students-give-them-extra-help-1527154200.

Berkeley Center for Teaching and Learning. 2018. *Remember and Understand. UC Berkeley: Division of Undergraduate Education*. Available at https://teaching.berkeley.edu/resources/engage/remember-and-understand.

Borzelleca, D. 2012. "The Male-Female Ratio in College." *Forbes*. Available at https://www.forbes.com/sites/ccap/2012/02/16/the-male-female-ratio-in-college/#18be4bf0fa52.

Bransberger, P. 2018. "Demographics, High School Graduates & Higher Education Demand." *Knocking at the College Door*. Available at https://static1.squarespace.com/static/57f269e19de4bb8a69b470ae/t/5b57353ff950b7329f3c9856/1532441920461/July+24+2018+PBransberger+Slides+for+web.pdf.

Briggs, S. 2016. "6 Ways Digital Media Impacts the Brain." *Open Colleges*. Available at https://www.opencolleges.edu.au/informed/features/5-ways-digital-media-impacts-brain/.

Campus Pride. 2018. "Colleges and Universities that Provide Gender-inclusive Housing." Available at https://www.campuspride.org/tpc/gender-inclusive-housing/.

Cervero, R. M. 1988. *Effective Continuing Education for Professionals*. San Francisco: Jossey-Bass.

Chawinga, W. D. 2017. "Taking Social Media to a University Classroom: Teaching and Learning Using Twitter and Blogs." *International Journal of Educational Technology in Higher Education* 14/3. Available at https://educationaltechnologyjournal.springeropen.com/articles/10.1186/s41239-017-0041-6.

Computer Business Review. 2018. "What is Google." Available at https://www.cbronline.com/what-is/what-is-google/.

Devlin, T. J., C. R. Feldhaus, and K. M. Bentrem. 2013. "The Evolving Classroom: A Study of Traditional and Technology-Based Instruction in a STEM Classroom." *Journal of Technology Education* 25/1: 34–54.

Eagan, K., E. B. Stolzenberg, A. K. Bates, M. C. Aragon, M. Ramirez Suchard, and C. Rios-Aguilar. 2015. "The American Freshman: National Norms Fall 2015." *Higher Education Research Institute*. Available at https://www.heri.ucla.edu/monographs/TheAmericanFreshman2015.pdf.

Facht, U. 2017. "The World's Largest Media Corporations, 2017." *Nordic Information Centre for Media and Communication Research*. Available at http://www.nordicom.gu.se/en/latest/news/worlds-largest-media-corporations-2017.

Galina, B. n.d. "Teaching First-Generation College Students." *Center for Teaching*. Available at https://cft.vanderbilt.edu/guides-sub-pages/teaching-first-generation-college-students/.

Gates, G. 2017. "In U.S., More Adults Identifying as LGTB." *Gallup: Social & Policy Issues*. Available at https://news.gallup.com/poll/201731/lgbt-identification-rises.aspx.

Gross, B. D. 2009. *Tools for Teaching*. San Francisco, CA: Wiley & Sons.

Heward, W. n.d. "Guided Notes–Improving the Effectiveness of Your Lectures." *Stanford University*. Available at https://tomprof.stanford.edu/posting/495.

Institute for Veterans and Military Families. 2017. "I am a Post-9/11 Student Veteran." *Student Veterans of America*. Available at https://ivmf.syracuse.edu/wp-content/uploads/2017/06/I-AM-A-POST-911-Student-Veteran-REPORT.pdf.

Jaschik, S. 2018. *2018 Survey of Faculty Attitudes on Technology: A Study by Inside Higher Ed and Gallup*. Available at https://www.insidehighered.com/system/files/media/IHE_2018_Survey_Faculty_Technology.pdf.

Jhally, S. 2011. "A Letter from Sut Jhally." *Media Education Foundation*. Available at http://www.mediaed.org/letter-from-sut-jhally/.

K-State. 2014. "The New Science of Learning: Concepts and Strategies for Changing the Way We Teach." Available at https://www.youtube.com/watch?v=P3IcDzr-tK4.

Linneman, J. 2018. "Share, Show, and Tell: Group Discussion or Simulations Versus Lecture Teaching Strategies in a Research Methods Course." *Teaching Sociology* 1/10. Available at https://journals.sagepub.com/doi/abs/10.1177/0092055X18799405.

McChesney, R. W. 2013. *Digital Disconnect: How Capitalism is Turning the Internet Against Democracy*. New York: The New Press.

McKeachie, W. and M. Synicki. 2013. *McKeachie's Teaching Tips*. Boston: Cengage Learning.

Moyers, B. 2017. "How Media Consolidation Threatens Democracy: 857 Channels (and Nothing on)." *Moyers*. Available at https://billmoyers.com/story/media-consolidation-should-anyone-care/.

National Center for Education Statistics. 2018. *Financial Aid. Fast Facts.* Available at https://nces.ed.gov/fastfacts/display.asp?id=31.

National Center for Education Statistics. 2016. "Total Fall Enrollment in Degree-Granting Postsecondary Institutions, by Attendance Status, Sex of Student, and Control of Institution: Selected Years, Selected Years, 1970 through 2026." Available at https://nces.ed.gov/programs/digest/d16/tables/dt16_303.40.asp?current=yes.

National Center for Education Statistics. 2016. "Status and Trends in the Education of Racial and Ethnic Groups 2016." Available at https://nces.ed.gov/pubs2016/2016007.pdf.

National Center for Education Statistics. 2017. "Total Fall Enrollment in Degree-Granting Postsecondary Institutions, by Attendance Status, Sex of Student, and Control of Institution: Selected Years, 1947 through 2027." Available at https://nces.ed.gov/programs/digest/d17/tables/dt17_303.10.asp.

Nichols, J. & McChesney, R. W. 2002. *Our Media, Not Theirs: The Democratic Struggle Against Corporate Media.* New York: Seven Stories.

Palis, A. G. & Quiros, P. A. 2014. "Adult Learning Principles and Presentation Pearls." *Middle East Afr J Ophthalmol* 21/2: 114–22.

Presnky, M. 2010. *Teaching Digital Natives.* Thousand Oaks: Corwin.

Sepulveda, J. 2016. "Introduction: Project-Based Learning." *University of Oregon.* Available at https://learn.canvas.net/courses/1123files/481829/download?verifier...wrap=1.

Shaw, A. 2009. "What is 21st Century Education? 21st Century Schools." Available at https:www.21stcenturyschools.com/What_is_21st_Century_Education.htm.

Shaw, C. M. 2017. "Using Facebook as an Educational Resource in the Classroom." *International Studies Association and Oxford University Press.* Available at http://oxfordre.com/internationalstudies/view/10.1093/acrefore/9780190846626.001.0001/acrefore-9780190846626-e-114.

Southern Connecticut State University. 2018. "First Year Experience (FYE) Program." Available at https://www.southernct.edu/offices/first-year-experience/.

Stanford University. 2018. "Project-Based Learning." *Stanford Teaching Common.* Available at https://teachingcommons.stanford.edu/resources/learning/learning-activities/project-based-learning.

Stoller, K. 2018. "The World's Largest Tech Companies 2018: Apple, Samsung Take Top Spots Again." *Forbes.* Available at https://www.forbes.com/sites/kristinstoller/2018/06/06/worlds-largest-tech-companies-2018-global-2000/#28d5b4e14de6.

Teniente-Matson, C. 2018. "Advice: 6 Steps to Help First Generation Students Succeed." *Higher Education* 65/26. Available at https://www-chronicle-com.scsu.idm.oclc.org/article/Advice-6-Steps-to-Help/242659.

The Center for Innovation in Teaching and Learning. 2018. *Lecture-Based Class*. Available at https://citl.illinois.edu/citl-101/teaching-learning/resources/teaching-in-specific-contexts/lecture-based-classes.

The Postsecondary National Policy Institute. 2016. "Factsheets: First-generation Students." Available at http://pnpi.org/first-generation-students/.

University of Michigan Center of Teaching and Learning. n.d. "Getting Started with Technology." Available at http://www.crlt.umich.edu/teaching-technology/getting-started.

University of Washington Center for Teaching and Learning. 2018. "Teaching with Technology." Available at http://www.washington.edu/teaching/teaching-resources/engaging-students-in-learning/teaching-with-technology-2/.

Wilkins, A. C. 2018. "New in Class: First-Generation Students and the Unseen Costs of College." *Phi Kappa Phi Forum*. Available at https://go.galegroup.com/ps/i.do?p=AONE&sw=w&u=googlescholar&v=2.1&it=r&id=GALE%7CA562377074&sid=classroomWidget&asid=632c45c1.

Zakrajsek, Todd. 2014. "K-State Teaching and Learning Center." Available at https://www.youtube.com/watch?v=P3IcDzr-tK4.

Chapter 9

Campfire, Curator, Deejay

Lectures and 'Lectronic Enhancement

Mike Mosher

There are various visual representations of the traditional classroom. The medieval scholar that is shown tutoring students in the early universities is shown as a robed savant, speaking to youthful protégés. The surgeon, as depicted by Rembrandt in 1632 or Thomas Eakins in the 1870s, demonstrates anatomy or procedures step-by-step in the medical college's operating theater, to a rapt audience, perhaps a horrified female family member too. Representing studio art courses, there are photographs of nineteenth-century ateliers, with the master painter surrounded in his studio by eager students who observantly draw from the live nude model (an essential part of the team). Add to that commercial visualizations of the university of "the digital community," the 1990s paradigm which has come of age to a significant degree, but not without its problems and contradictions.

A lecture is a form of oratory, a forensic with a long tradition. In the United States, the public gathered to hear visiting lecturers in Chautauquas—originally in Chautauqua, NY—and eminent religious leaders were noted for their oratorical excellence. Ministers preaching in African American churches in our time are notable for effective use of emphasis, repetition, cadence. I don't know if I am influenced by the few times I listened to a bit of Rev. C. L. Franklin on Detroit radio in the 1960s but I sometimes jokingly finish a class lesson saying "end of sermon" or "The Mass is ended, go in peace."

I have not followed a traditional academic career, and I am thankful that being a rebellious faculty brat, going against family expectations, gave me decades of various creative and professional experiences that I now bring to the classroom. Some of my "campfire" stories are of mistakes I made in corporate settings, hard won lessons. My own career outside of academia has been based on acquisition of various skills and expertise, making me now an advocate to my students of fearless hybridity. The satisfaction of my dual appointment, teaching pencil, ink and paint in the Art Department, and teaching design skills and visual culture in the Communication Media Administration (CMA) graduate program, affirms John Naisbitt's concept of "High tech and high touch," articulated in his 1982 book *Megatrends* where increased numbers of jobs involving information processing (CMA) is compensated by increased sensual pleasures (Art).[1] I witnessed that dichotomy balanced in software engineers being connoisseurs of wineries and bicycling in Silicon Valley. The most compelling lectures I ever saw were the computer scientist-theorist Ted Nelson's mesmerizing autobiographical talks in Palo Alto circa 1991; as each wrapped up, I was amazed that three hours had passed.

Grant Wiggins observed, "It's not teaching that causes learning. Attempts by the learner to perform cause learning, dependent upon the quality of feedback and opportunities to use it."[2] The professor's job to challenge, to broaden the scope, to bring understanding and appreciation, and to say: "Life is sweeter, fuller if you know about this. You will have more understanding, power and flexibility."

The lecture is paramount to me, despite a further complication. It is always an effort to speak loudly because I have a damaged voice. Like George Orwell, I have a single functioning vocal cord, which gives me a very soft, gruff voice, so I am aware of its limitations no matter how loquacious I am during that class session.[3] Consequently, I work to modulate it, even exaggeratedly, to avoid a raspy monotone and keep students' attention.

I know the importance of eye contact, and direct each sentence or two to a different student. Eye contact individualizes the experience, reminds and reinforces that "he's talking to me." Of his friend Walter Benjamin, Gershom Scholem recalled sixty-eight years later:

> Before I made Walter Benjamin's personal acquaintance, I saw him in the autumn of 1913 at a meeting that took place in a hall above the Cafe Tiergarten in Berlin . . . the main spokesman of [one group] was Walter Benjamin, who was rumored to be their most gifted intellect. He made a very tortuous speech . . . . I cannot recall the details of the talk, but I shall never forget his manner of presentation. Without looking at the audience, he delivered his absolutely letter-perfect speech with great intensity to an upper corner of the ceiling, at which he stared the whole time.[4]

Personal eye contact is effective in a classroom lecture if the instructor is addressing about two dozen individuals or less, but above that, might it as well be recorded.

What follows are largely contemplative reflections on my classroom experience rather than research. I shall introduce three metaphors for the contemporary classroom and its instructor: campfire, curator, and deejay.

THE METAPHORS

Around the campfire the instructor imparts not only data and information, but hopefully knowledge (information organized) and wisdom (knowledge proven over time). In the drawing class, I often talk about aspects of our class work that I struggled with in college, and acknowledge the difficulty of the skills instruction, and assignments testing it, in which they are endeavoring competence. To extend the rustic metaphor, I am the trail guide leading them through the dangerous woods of drawing. The campfire brings context to the information and the lore the storyteller conveys.

The curator is a connoisseur, an educated sensibility, sharing her or his knowledge and expertise, bringing form to the content. A museum, gallery, or exhibition is assembled to present multiple works coherently, in a way that showcases an important thesis, theme, or themes.

A deejay is a disk jockey of instruction, propelling the party as a seamless, satisfying experience from start to finish, knowing when to turn up the volume, when to change the mood, attentive to providing a memorable duration. Like a theatrical or musical performance, content over time must be engaging or lose the heart of the crowd. She or he is a curator, but one of a timed experience. Learning is the dance. She is concerned with Bergson's distinction of *durée* (duration) rather than merely *temps* (moment).[5] As David Bowie sang, "I am a D.J, I am what I play . . . I've got believers, believing in me."[6] The instruction is internalized and manifested in the performer, the lecturer.

MIXED METAPHORS IN CLASSES

My regularly scheduled classes enjoy different degrees of technological enhancement, which I will express in the above metaphors. I've developed a mix of spoken lecture and electronics—online teaching systems, YouTube videos, and videoed lectures—for classes that are both theory and skills based, with the goal of all classes maximally pedagogical yet not pedantic. But regardless of the class, what is important is that the instructor demonstrates the skills with the assigned tools around the campfire to establish the instructor's credibility.

Beginning Art Class

The beginning drawing class is almost entirely hands on, where, after a campfire demonstration of drawing skills, students draw the day's still life, model, or other assignment. Each class ends with group critique of the day's work. A drawing, painting, or graphics software demo is a lecture, but with its argument, its endpoint taking shape before the students' eyes. Each of the six units (still life, hatching/modeling value, the skeleton, the figure, one-point perspective, two-point perspective, pen, and ink) begins with a lecture demo, usually about ten minutes long, with students then drawing until the daily half-hour, participatory, group critique.

The online learning system is used in this class to post the syllabus, some curated instructional examples, and relevant examples from art history or sketchbooks. When these were passed around the class in physical books, students barely glanced. When the instructional examples were distributed as paper handouts, many were left in the classroom.

The campfire flares up when an ad-hoc momentary lecture fragment is spurred when the instructor makes the rounds of the easels and sees problems in one student's work: "Everybody, don't forget to draw a 'cloud' incorporating all objects on the table at the start of your drawing." For the reason that an individual work might make me begin to speak to them collectively, I jocularly tell the class not wear earbuds, for to do so might make you miss this comment, inspired by a single student but directed to the class.

The last half-hour of each 170-minute class session, the drawing critique, is directed and active, the instructor asking each of the students to evaluate their peers' work, criticize as well as recognize strengths, based upon the day's shared drawing experience.

Art Community Mural Class

The Art Community Murals class includes several lectures projecting historical imagery—listen around the campfire to the lore of this medium, from Mexicans in the 1930s through contemporary Detroiters—as well as personally demonstrating painting and digital design skills. Besides viewing imagery and video links loaded on Canvas, students post selfies taken in the library at favorite images in its massive *Murals of Diego Rivera* book, describing what the mural in the book teaches them that's applicable to the semester's public mural projects.

Comic Class

In the classroom it is taught with a mix of historical lectures, physical drawing and design demonstrations, videos, and other resources for students to access on the online learning system, plus books and comic books that have been discussed passed around in class—around the campfire—for them to look at the artwork within hands on. The lectures offer a basic literacy in over 120 years of American comics; a third of the course content is thus art history. Classes have designed and drawn their final comics from the scripts provided by English classes. Students have provided spot illustrations for math story problems and for writing center student memoirs. Like the other classes, students have time to work on their assignments in class, where the roving professor critiques their progress.

Visual Criticism Class

The graduate program's visual criticism class uses a traditional textbook[7] as well as campfire and curated elements supplementary readings, plus lectures that are the professor's gloss upon the wide-ranging content, often including examples from his own art and design practice. There are also YouTube videos, especially several of John Berger's 1972 *Ways of Seeing* BBC television series. Student discussions are often difficult to grade because, for the majority of the class (often 40% Saudi and 40% Chinese), English is the second language.

I am very much the deejay as well, for electronic enhancements such as YouTube, or JPEG imagery, or links to readings lets the instructor supplement the lecture, but still respects linearity, her or his control of duration from start to finish. And I probably discovered long after my students did that a lecture video on YouTube can be watched at 1.25 or 1.5 times the speed at which it was recorded, with no "chipmunk" rise in the speaker's pitch. I always list a video's run time when I post it in Canvas, but students probably take less time to watch them.

Design Class

In this class, I light the campfire when I offer up stories about my experiences in graphics and media development in the Silicon Valley tech industry. I frequently tell stories of mistakes I made, and what I learned the hard way. A student once negatively commented on a class, "He just tells us about how things were done in big California companies, then about new technology that excites him." But it is important to show students not only the current tools and design skills but also the historical and cultural context so they can succeed.

Multimedia Tools Class

This class is intended for media managers, requiring of them enough hands-on experience to understand the issues the vendors (illustrators, photographers, designers) will work with and to effectively communicated with them. In this manner students also gain the understanding to argue with their own management for sufficient resources for high-quality results. Here I teach some design tools whose fundamentals haven't changed in a quarter-century, only their specifics. YouTube instructional videos might excellently elucidate a special process, yet the tools menu shown onscreen might look a bit different now than five years ago when the video was made. I believe any course teaching software should most of all teach students how to learn software, for it will continually change and there will always be new creative applications on the market.

User Experience Class

This is a graduate course that had been entirely project-based for several years, where students prototype a variety of screen and device interfaces for websites, phone apps, a 3D or VR instructional game, and future technology. Before each unit I give a lecture, my favorite being the advance of graphic user interfaces and technology in the past 120 years, especially since the Macintosh, a welcome facilitator of my own Silicon Valley career. Yet as the graduate program's emphasis has shifted from multimedia production to media administration, students now also give oral or written presentations on contemporary media producers they have researched, and brief introductory "elevator pitches" to potential employers in the tech industry. It is hoped that the class and program prepare them to "follow their noses" and look out for opportunities in new, ever-changing technologies.

Online Classes

These classes have short historical and skill-demonstrating lectures, plus readings as documents or links, posted to the online system. The students' assignments are then uploaded to discussions, so their classmates can critique prior to my grading and feedback. An online class demands much more curation, and the deejay's knack of keeping students' attention from a distance. To teach online requires students to watch "talking head" lectures by me seated at my Macintosh's camera, but each lecture is intercut with imagery by the artists discussed, before it is posted on YouTube.

Economy is valued in online lectures. At Gettysburg battlefield in April 1863, a long and long-anticipated speech was delivered by celebrated orator Edward Everett, followed by a short speech by President Abraham Lincoln, yet Lincoln's is most remembered and memorized in American grade schools at least a century after. For online classes at my university, we have been advised to break video lectures for online instruction into segments of about 15 minutes, and I am surprised at how my previously 45-minute lectures break down effectively into two 15-minute ones or even a single 15 minute one. When online, students can watch these lectures repeatedly as necessary.

Per the suggestion of the best student in past courses attendance was required at either one of two real-time chat sessions, if only for the student to say, "Hi, Comic penciling going well, will post it Sunday. No questions about the reading. On my break at grocery store, see ya." Chat communications in my online class may be on a Saturday morning, answering students' questions—some from students in Taiwan or central China. Online, I have wondered: should I substitute two-way audio or video for text chat? I don't really see an advantage, especially with the speed of students' texting. The chat function is like the classroom, in that all students present see the professor's answer to every student question.

Sometimes students reply "LOL" to the professor's mild joke or pun. Does a lecture and its feedback offer emotional benefit for the professor? Otherwise, she or he is community-less producer, a production unit of alienated labor. There is, granted, a certain loneliness in teaching online. Am I really teaching, reaching them? Their successful artwork, and strong grades in their quizzes, make me think so but the lack of eye contact generally keeps me seeing the light bulb of understanding go off, the spark of epiphany in their eyes, affirming the moment a lesson take root, even tentatively, then blossoms (as over subsequent drawings). I get that in a lecture.

ENGAGEMENT AND DISTRACTION

Claire Howell Major writes:

> It can be challenging for students to maintain full attention during long lectures; students may not be effective active listeners, or well-developed note-takers; and students can be tempted to task switch or "multitask."[8]

Cindy Finelli writes that passive student resistance takes forms of "working on unrelated classwork or pretending to participate."[9] Major advocates segmented lectures be punctuated with selected "active learning" methods, opportunities to apply the content in discussion, problem-solving exercise, peer teaching, or quizzing each other. She advises lectures be focused, in a clear and initially explained format, with supportive audiovisuals in whatever form. Greater student involvement requires a supportive learning environment, mutual respect, ground rules for participation; the instructor gauges student engagement from prompt arrival and readiness to participation and body language; and students' ability to link new information to early concepts and methods.[10]

My use of online systems to supplement lectures is not necessarily active learning: it's most often organizing and depositing essential and supportive content into pigeonholes, to be choreographed by me in class, or experienced by student on her or his own time. Yet it is active when students use it to post critiques of classmate's work, take a quiz, or to chat.

And there are 800-lb. gorillas in our classroom, residing in every bookbag or pocket. The Android or iOS phone is certainly the enemy of the lecture, perhaps the lecture's very opposite. I was probably the last person in Michigan to buy an app phone, which is ironic as my classes for media managers endeavor to provide students skills to manage app production. On campus, whether the student is in class or walking, the phone distracts them rather present a specific, cogent part of it (a lecture). It is designed to demand immediate attention with messenger's ping or ever-changing Facebook or Twitter experiences. It fights the location-based lecture as it simultaneously fights linearity, built-up argument, and content

I tell my class to put away their devices and laptops—international students, however, especially Chinese ones, used to like to quickly look up words I use in their online dictionaries—for I too know somewhere, that something more interesting than classwork is going on. There is the oft-irresistible lure of being someplace else. Facebook is a marvelous escape, which I compare to an Irish pub: I amble in, learn of someone's new baby or graduation, argue politics, make a quip and laugh at others', and return to work.

CONCLUSION

The traditional university lecture is analogous to the cinema or theater performance, as a collective experience where audience attention is directed to the delivery up front, at a specific time. Tape, disc, and streaming distribution has changed cinema, and offer a private and individual time-shifted experience too. I've similarly seen the atmosphere in exuberant bars changed with the addition of television screens on multiple sports, where, rather than stimulating group commentary (and argument), patrons are lost in silent viewing and contemplation.

I believe art and design professors need to continually show students and the campus community that their practice addresses the media they teach and that they know what they are talking about. As we grow in experience over time, we hope to increasingly take our teaching to the higher level: from disconnected data to information, from information to knowledge, and from knowledge to wisdom. The contemporary professor stokes a campfire, where students are gathered around to hear stories of professional experience that, one hopes, brings wisdom as well as knowledge, see skills demonstrated or learn of the information in books. That instructor is also a curator who assembles resources, organizes them into a clearly labeled museum of instructional content. And finally, an educator is a deejay, presenting the most engaging and memorable hits, productions, and occasionally obscurities of the subject being taught, for retention and forward movement.

ONLINE RESOURCES

1. The medieval classroom, discussed in Lost Kingdom
2. Rembrandt 1632 painting "The Anatomy Lesson of Dr. Tulp"
3. Thomas Eakins' 1875 painting of a professor of surgery lecturing "The Gross Clinic"
4. Thomas Eakins' 1889 painting of a professor surgery lecturing "The Agnew Clinic"
5. Apple's 1994 internet service eWorld, with its Learning Center, is discussed in Macworld
6. Encyclopedia Britannica on the Chautauqua Movement
7. "The Inner Conflict" sermon (1975) by Rev. C.L Franklin
8. John Naisbitt *Megatrends* (1982) reviewed by John T. Slantia, "High Tech/High Touch." *Bookpage*, October 1999
9. Ted Nelson, "Here I Stand, at Age 80" (2017)
10. Gershom Scholem, "My Friend Walter Benjamin." *Commentary*, December 1, 1981
11. Henri Bergson, *DURÉE ET SIMULTANÉITÉ: À propos de la théorie d'Einstein* (1922)
12. David Bowie, video of his song "D.J."
13. SVSU Art 433 students' community murals
14. SVSU Art 390 students' comics
15. John Berger, "Ways of Seeing," BBC television series (1972)
16. President Abraham Lincoln's Gettysburg Address, at Encyclopedia Britannica
17. Cindy Finelli quoted in Nicole Casal Moore, "Active Learning Effort a Step Toward Broader Change in Higher Ed." *The University Record*, October 8, 2018

NOTES

1. John T. Slantia, "High Tech/High Touch." *Bookpage*, October 1999. Available at https://bookpage.com/reviews/779-john-naisbitt-high-tech-high-touch#.W7lVSFJRf6A.

2. Leslie Richardson, "Turning Your Classroom Inside Out." *NEA Higher Education Advocate* 29/3 (May 2012): 6–9.

3. David Smith and Mike Mosher, *George Orwell Illustrated* (Chicago: Haymarket Books, 2018), 138.

4. Gershom Scholem, "My Friend Walter Benjamin." *Commentary*, December 1, 1981. Available at https://www.commentarymagazine.com/articles/my-friend-walter-benjamin/.

5. Henri Bergson, *DURÉE ET SIMULTANÉITÉ: À propos de la théorie d'Einstein* (1922). Available at http://classiques.uqac.ca/classiques/bergson_henri/duree_simultaneite/duree_et_simultaneite.pdf.

6. David Bowie, "D.J." July 4, 2018. Video available at https://www.youtube.com/watch?v=MRRmU_pOXnk.

7. Marita Sturken and Lisa Cartwright, *Practices of Looking* (Oxford: Oxford University Press, 2017).

8. Claire Howell Major, "Engaging Students through Interactive Lecturing." *NEA Higher Education Advocate* 36/5 (September 2018): 6.

9. Nicole Casal Moore, "Active Learning Effort a Step Toward Broader Change in Higher Ed." *The University Record*, October 8, 2018. Available at https://record.umich.edu/articles/active-learning-effort-step-toward-broader-change-higher-ed.

10. Major, "Engaging Students," 6.

WORKS CITED

Bergson, Henri. 1922. *DURÉE ET SIMULTANÉITÉ: À propos de la théorie d'Einstein*. Available at http://classiques.uqac.ca/classiques/bergson_henri/duree_simultaneite/duree_et_simultaneite.pdf.

Bowie, David. 2018. July 4. "D.J." Video available at https://www.youtube.com/watch?v=MRRmU_pOXnk.

Major, Claire Howell. 2018. "Engaging Students through Interactive Lecturing." *NEA Higher Education Advocate* 36/5 (September 5): 6–9.

Moore, Nicole Casal. 2018. "Active Learning Effort a Step Toward Broader Change in Higher Ed." *The University Record*, October 8. Available at https://record.umich.edu/articles/active-learning-effort-step-toward-broader-change-higher-ed.

Richard, Leslie. 2012. "Turning Your Classroom Inside Out." *NEA Higher Education Advocate* 29/3: 6–9.

Scholem, Gershom. 1981. "My Friend Walter Benjamin." *Commentary*, December 1. Available at https://www.commentarymagazine.com/articles/my-friend-walter-benjamin/.

Slantia, John T. 1999. "High Tech/High Touch." *Bookpage*, October. Available at https://bookpage.com/reviews/779-john-naisbitt-high-tech-high-touch#.W7lVSFJRf6A.

Smith, David and Mike Mosher. 2018. *George Orwell Illustrated*. Chicago: Haymarket Books.

Sturken, Marita and Lisa Cartwright. 2017. *Practices of Looking*. Oxford: Oxford University Press.

Chapter 10

The Lecture in Political Science Education

Unpacking a Paradox

John Craig

There is much that has changed in political science education in universities over the last 100 years.[1] The curriculum has been influenced by a range of intellectual movements such as behavioralism, feminism, and rational choice. In many countries, the development of the taught discipline has been shaped by revolutions, wars, and changing expectations of the scope and role of higher education. There have been waves of democratization, decolonization, and globalization that have shaped both what is studied and the context in which it is taught. There have also been changes to the student body, with large scale expansions of university systems in many parts of the world, and new developments in the technology that is available for teaching.

Throughout all this change the role of the lecture has been remarkably constant. In the overview report for an international survey of political science undertaken by UNESCO in the early 1950s, W. A. Robson could confidently report that "the lecture course is a universal method which undoubtedly forms the principal instrument for teaching political science."[2] Subsequent surveys conducted in a number of countries over the last sixty years have shown that this continues to be the case.[3] Many would agree with Jeffrey Bernstein's observation that "virtually every professor teaching political science" uses lectures and that "the vast majority of students who take political science at college or university level experience it as predominantly a lecture orientated field."[4]

The predominance of a particular form of teaching might not be of anything more than passing interest where it not for the extent to which this had been subject to long-standing criticism from both inside and outside the discipline. Educationalists have long questioned the value of the lecture. Writing at the start of the 1970s, Donald Blight reviewed the results of nearly a hundred studies and concluded that while the evidence suggested that lectures were no better or worse than other means of transmitting information, they were less effective at developing students as independent thinkers or changing their attitudes.[5] Blight's work has continued to be a point of reference in many books written for college and university teachers which highlight the limitations of the lecture.[6] These identify lectures as a teaching method based on the transmission of knowledge to a passive audience. Such teaching, it is argued, tends to be ineffective as students will lose concentration and retain only a little of what has been said.

There were also prominent voices within the discipline that questioned the predominance of the lecture. Writing in the *American Political Science Review* in 1940, Walter Laves, then chair of the Department of Political Science at the University of Chicago referred to how an "increasing number of us realise the very limited value of lectures."[7] When an American Political Science Association Committee for the Advancement of Teaching reported in 1951, it noted that lectures appeared with "monotonous regularity" in accounts of teaching methods, and called for the discipline to "vitalize and modernize" its approach to education.[8] While noting their predominance, W. A. Robson had also questioned the value of the lecture, referring to it as a "medieval device which evolved at a time when the ability to read was a rare accomplishment." He argued that its overuse as a method for communicating information, that could just as well be read, was wasteful of time and likely to result in students becoming "passive, bored, indifferent, and unreceptive."[9] Such criticisms are regularly voiced in the steadily growing literature on teaching in political science and have prompted the development of a range of pedagogic innovations such as simulations, internships, civic engagement, and the use of film and literature.

There is a paradox then. While the limitations of the lecture as a tool for teaching political science has been acknowledged for at least sixty years, from both inside and outside the discipline, it has continued to be the most widely used type of teaching. There are a range of explanations that have been put forward for why this might be the case.[10] Some point to the architecture of the university, with its ready supply of lecture theatres, and see this as a determining influence on the types of teaching that can take place. Others point to the "cost effectiveness" of the lecture, providing a relatively cheap method of teaching large numbers of students. There are also those who argue that the continued use of the lecture is a response to student expectations. This may be the result of traditional images of the university, or consumerist demands that the course content is packaged and delivered in readily accessible chunks.

Finally, there are those who look for explanations in the values and attitudes of the academic profession. Political science, it has been suggested, is a conservative discipline, reluctant to develop new methods of teaching, even if the old are shown to be ineffective.[11] As Robert Glover has put it in the introduction to a recent book on teaching politics, "Though higher education is often cast as the wellspring of society's most innovative thinkers and problem solvers, it can, in its approach to teaching and learning, be remarkably insular and closed minded."[12]

As is often the case, there is a certain amount of truth in each of these explanations. Any individual academic joining a university and finding their classes scheduled in large tiered lecture theatres is no doubt constrained in the choice of teaching techniques they can adopt. Equally, a university administrator, who has to plan the delivery of a course within a given set of resources may decide that large lectures are at least part of the solution. Indeed, the use of lectures in one part of a course, may allow greater scope for more intensive small group or individual tuition elsewhere. There may also be some tendency within the political science profession, as there are in many others, to a conservative approach to practice. If someone has succeeded in learning through a particular method, it would not be surprising if this is the way they chose to instruct others. However, these reasons do not seem sufficient to account for such widespread use of the lecture over such a long period of time. They are all deficit explanations, based on limitations, failure, or scarcity of resource in particular circumstances. What they lack is any positive reasons why academics have chosen to continue to make extensive use of the lecture over many decades and locations as one of their preferred methods of teaching political science.

In a book written for students of the social sciences and humanities, the political scientist Patrick Dunleavy, outlined four positive reasons why academics might choose to teach through a lecture.[13] First, the lecture could be used to provide an overview of a topic for students. Critics might respond that this could equally be secured through independent study or other types of learning, but this is answered in part by Dunleavy's second and third reasons. Through the lecture, a teacher can introduce students to material in a more accessible form than it might be presented in the literature and can cover recent ideas, issues, events, and debates that have not yet reached publication. This is particularly important in political science. As well as the study of theoretical and historical material, it also involves the study of contemporary political relationships, which are the subject of continual and sometimes dramatic change.[14] As events unfold, it is likely that students and teachers will want to explore these long before they are covered in books or journal articles, and lectures are one way in which this can be done.

Dunleavy's fourth rationale for lecturing is that it provides students with an opportunity to see how an academic argument can be built, and he refers to the benefits of seeing such arguments being made in person.[15] It is a factor that should not be overlooked. In other walks of life, such as sport, music, and theatre, people are often willing to put additional time and resource into experiencing a live performance when they could more easily access a recording. Again, the context of political science is relevant. The making of arguments is an integral part of the discipline, and through engaging with lectures, students can observe how this is done. This, it should be noted, requires something more than a passive process of absorbing knowledge. Students will get the most from a lecture if they are able to follow the development of the argument that is presented, consider how evidence has been selected and used, and compare the content and styles of different teachers. Students will often need to be supported in developing such critical listening skills and have been provided with advice on how to prepare for the session. This might include reading around the topic or studying questions of method.

The point has also been made that not all teachers are inspirational presenters, and some may even have dull lecturing styles.[16] However, the contrast between these two extremes can be unhelpful. It would be unrealistic to expect that every lecture was inspirational, but there is also no evidence that most lectures are poorly delivered. We might expect that most are somewhere in between competently presented and reasonably interesting. In the current environment, it is common for academics to receive feedback from students on their classes, creating regular opportunities for them to reflect on how aspects of content or delivery might be enhanced. Likewise, there are many resources available that provide advice to academics on how they can improve their lectures.[17] The point should also be made that any comparison between lectures and other types of teaching, also relies on the effectiveness of the available alternatives and the ability of teachers to implement these.

While these aspects are important in explaining the continued use of the lecture, they need to be complemented with a range of perspectives which identify how the lecture has developed in recent decades. These fall into three categories. First, there are a number of approaches that aim to enhance the lecture through improving it as a means of communication. Second, there are approaches that explore how lectures can be made more interactive to help students engage with the material and create dialogue within the classroom. Third, there are approaches that combine the use of lectures with other approaches to teaching, recognizing their specific usefulness in certain contexts.

Some have approached the paradox of the lecture by trying to improve the way in which it is delivered. It is easy to take for granted the availability of computers, data projectors, and software packages such as PowerPoint and Prezi. These may not always be used as effectively as they might, but they have opened up a range of ways in which the delivery of a lecture can be enhanced. These include the use of charts, pictures, and graphs, and the integration of short video clips.[18] Such techniques extend the opportunities to engage students visually, and animated graphics can help teachers to illustrate dynamic relationships and change over time. In addition, some packages allow teachers to present virtual mind maps which can be used to "demonstrate spatial relationships between theories, concepts, ideas, and events."[19]

There are also approaches that use technology to provide new opportunities for students to engage with the lecture. The use of such flipped or inverted classroom approaches, which sometimes use online lectures to prepare students for face-to-face sessions, has grown in recent years and positive results are reported when they have been used in political science.[20] Educators have also experimented with the use of podcasts and mini video lectures to support students to consolidate or develop their learning in combination with other methods.[21] Use has also been made of lecture capture systems, which allow students to experience the live lecture, and later access a recording.[22] The opportunity this provides for students to revisit a lecture, can take the pressure away from note-taking during the class and facilitate more active and critical listening.

Others have tried to make the lecture more interactive. This can be achieved in a number of ways, such as pausing the lecture and breaking students up into small "buzz groups" to discuss a topic or the answer to a question.[23] For example, question-based outlines involve asking students questions as a device for eliciting responses and establishing a dialogue in the classroom.[24] Such approaches can also be used to help students identify what they already know about a topic at the beginning of a lecture, which can facilitate a more critical and engaged approach.

Technology can also assist in making lectures more interactive through audience response systems often known as clickers. Students are provided with access to a system (through either a device of their own or one that is distributed in class) that can collect and collate their responses to multiple choice questions in real time.[25] The use of these response systems has a number of benefits. They can provide an opportunity for interaction between the students and the lecturer, even in a large class. Students can express a view, see it represented, and responded to, which can mitigate some of the passivity of the traditional lecture. In addition, it can provide useful feedback to the teacher. This might be used to help them assess students' understanding of a topic at the start of a class, or to check how well they have understood content during the course of the lecture. This can help them to decide if they need to provide further explanation of a point, perhaps by using a different example or way of presenting it, or to move forward in the knowledge that students have understood the topic.

To some extent, the same types of interaction could be achieved without the use of technology. Students could simply be asked to raise their hand in response to a question. Apart from the logistical advantages of the clickers (which can count the responses of large numbers of students more quickly and accurately than a lecturer), they also provide anonymity to the individual student. Students may not always feel confident in expressing their views in a large group, and clickers provide an opportunity for them to do so more privately.[26] Overall, the evidence suggests that such systems can be used to increase both student learning and satisfaction.

The literature also provides many examples of how lectures can be most effectively combined with other methods of teaching. This is a key point for although lectures have been the most widely used form of teaching in politics, the evidence also suggests that they have not usually been used in isolation. Lectures have been found to be "an extremely effective means of teaching students in political science and international relations" when used in combination with discussion classes.[27] There is also evidence that lectures are better at achieving some outcomes than some other methods of teaching. One study which compared lectures and in-class debates found that while debates were "more effective in developing students' comprehension of complex concepts and application and critical evaluation skills," lectures appeared to be better "for promoting basic knowledge of the subject material."[28]

In addition, in contrast to approaches that pit lectures against more innovative approaches to teaching, lectures can be an integral element of how such innovations are delivered. They can, for example, be used to communicate essential background information or instructions which are needed for a simulation.[29] Likewise, in cases where films have been used in teaching, lectures have been used to introduce students to the topics and the background to the film.[30] In these cases, lectures appear to be a trusted technique for ensuring that students have access to essential information that is bespoke to the teaching context and lays the foundation for developing further engagement through more interactive teaching methods.

This chapter has sought to explain the paradox of the lecture in political science and account for how a form of teaching, that has been so long criticized from inside and outside the discipline, has remained so widely used. While there are a range of negative explanations for its continued use, this chapter has identified some of the positive reasons that educators choose to use lectures. While some of these would be applicable to a range of disciplines, others are specific to the challenges of teaching political science. This chapter has also examined some of the ways in which teachers of political science are developing their use of the lecture, in some cases through the use of technology and in others through combinations with other methods of teaching. As such, the lecture is likely to continue to be widely used by political science educators for the foreseeable future.

NOTES

1. On factors influencing the development of political science as an academic discipline see: J. Farr and R. Seidelman, eds., *Discipline and History: Political Science in the United States* (Ann Arbor: University of Michigan Press, 1993); D. Easton, J.G. Gunnell, and M.B. Stein, eds., *Regime and Discipline: Democracy and the Development of Political Science* (Ann Arbor: University of Michigan Press, 2017).

2. W.A. Robson, *The University Teaching of Social Sciences: Political Science* (Paris: UNESCO, 1954), 72.

3. N. Stammers, H. Dittmar, and J. Henney, "Teaching and Learning Politics: A Survey of Practices and Change in UK Universities," *Political Studies* 47/1 (1999): 114–26; S.G. Hartlaub and F.A. Lancaster, "Teacher Characteristics and Pedagogy in Political Science," *Journal of Political Science Education* 4/4 (2008): 377–39; M.J. Moore, "How (and What) do Political Theorists Teach: Results of a National Survey," *Journal of Political Science Education* 7/1 (2011): 95–128; C. Harris and B. Quinn, "Teaching, Assessment and Professional Development: Praxis in Ireland's Political Science Community," *Irish Political Studies* 32/2 (2015): 255–75.

4. J. Bernstein, "Signature Pedagogies in Political Science: Teaching Students How Political Actors Behave," in *Exploring More Signature Pedagogies: Approaches to Teaching Disciplinary Habits of Mind*, N.L. Chick, A. Haynie, and R.A.R. Gurung, eds. (Sterling, VA: Stylus, 2012), 90.

5. D.A. Blight, *What's the Use of Lectures?* (Harmondsworth: Penguin, 1972).

6. W.J. McKechnie, *McKechnie's Teaching Tips: Strategies, Research, and Theory for college and University Teachers* (Boston: Houghton Mifflin Company, 1999); J. Biggs and C. Tang *Teaching for Quality Learning at University* (Maidenhead: McGraw Hill, 2007).

7. W. Laves, "The Next Decade in Political Science Teaching," *American Political Science Review* 34/95 (1940): 985.

8. American Political Science Association, *Goals for Political Science: Report of the Committee for the Advancement of Teaching, American Political Science Association* (New York: William Sloane Associates, 1951), 277–8.

9. Robson, *The University Teaching of Social Sciences*, 73.

10. See for example D. Laurillard, *Rethinking University Teaching: A Conversational Framework for the Effective Use of Learning Technologies* (New York: Routledge Falmer, 2002): 92–4; S. Brown and P. Race, *Lecturing a Practical Guide* (London: Kogan Page, 2002): 21–61.

11. A. Leftwich, "Room for Manoeuvre: A Report on Experiments in Alternative Teaching and Learning Methods in Politics," *Studies in Higher Education* 12/3 (1987): 311–23.

12. R.W. Glover, "Introduction: 'Like captured Fireflies . . .'" in *Teaching Politics Beyond the Classroom*, R. Glover and D Tagliarina, eds. (New York: Bloomsbury, 2013), 2.

13. P. Dunleavy, *Studying for a Degree in the Humanities and Social Sciences* (Basingstoke: Macmillan, 1986), 10–11.

14. On this point see J. Craig, "What (if Anything) is Different about Teaching and Learning in Politics?" in *Teaching Politics and International Relations*, C. Gormley-Heenan and S. Lightfoot, eds. (Basingstoke: Palgrave Macmillan, 2012): 31–34.

15. Dunleavy, *Studying for a Degree in the Humanities and Social Sciences.*

16. See Robson, *The University Teaching of Social Sciences*, 72–86; Blight, *What's the Use of Lectures*, 39–40; Laurillard, *Rethinking University Teaching*, 92–4.

17. See the resource list that accompanies this chapter.

18. R.E. Hassner, "Sliding into Home Plate: How to Use Slideware to Improve Your Presentation (While Dodging the Bullets)," *PS: Political Science and Politics* 38/3 (2005): 393–97; M.L. Schrad, "In Defense of the Populist Lecture," *PS: Political Science and Politics* 43/4 (2010): 759–65.

19. E. Mayhew, "The New Generation of Prezi Presentation Software, Provided by Prezi Inc., and Student Engagement and Learning Within Political Science," *Journal of Political Science Education*, Published online: November 8, 2018, DOI: https://doi.org/10.1080/15512169.2018.1485572.

20. M. Touchton, "Flipping the Classroom and Student Performance in Advanced Statistics: Evidence from a Quasi-Experiment," *Journal of Political Science Education* 11/1 (2015): 28–44; S. Jenkins, "Flipping the Introductory American Politics Class: Student Perceptions of the Flipped Classroom," *PS: Political Science and Politics* 48/4 (2015): 607–11.

21. P. Woodcock and G. Duckworth, "Ipod Therefore I Am: Using PC Videos to Aid the Teaching of the History of Political Philosophy," *European Political Science* 9/1 (2010): 25–33.

22. D. Mallinson and Z. Baumann, "Lights, Camera, Learn: Understanding the Role of Lecture Capture in Undergraduate Education," *PS: Political Science and Politics* 48/3 (2015): 478–82; J.C. Roberts, "Evaluating the Effectiveness of Lecture Capture: Lessons Learned from an Undergraduate Political Research Class," *Journal of Political Science Education* 11/1 (2015): 45–60.

23. C. Leston-Bandeira, "Enhancing Politics Teaching through Active Learning," in *Teaching Politics and International Relations,* Cathy Gormley-Heenan and Simon Lightfoot, eds. (Basingstoke: Palgrave Macmillan).

24. J.C. Huerta, "Getting Active in the Large Lecture," *Journal of Political Science Education* 3/3 (2007): 237–49.

25. M. Velasco and G. Çavdar, "Teaching Large Classes with Clickers: Results From a Teaching Experiment in Comparative Politics," *PS: Political Science and Politics* 46/4 (2013): 823–29; Z.D. Baumann, K. Marchetti, and B. Soltoff, "What's the Payoff?: Assessing the Efficacy of Student Response Systems," *Journal of Political Science Education* 11/3 (2015): 249–63.

26. L. Holland, P. Schwartz-Shea, and J.M.J. Yim, "Adapting Clicker Technology to Diversity Courses: New Research Insights," *Journal of Political Science Education* 9/3 (2013): 273–91.

27. M. Krain and J.S. Lantis, "Building Knowledge? Evaluating the Effectiveness of the Global Problems Summit Simulation," *International Studies Perspectives* 7/4 (2006): 395–407.

28. M.Y. Omelicheva and O. Avdeyeva, "Teaching with Lecture or Debate? Testing the Effectiveness of Traditional Versus Active Learning Methods of Instruction," *PS: Political Science and Politics* 41/3 (2008): 603–7.

29. B.K. Dougherty, "Byzantine Politics: Using Simulations to Make Sense of the Middle East," *PS: Political Science and Politics* 36/2 (2003): 239–44; P.S. Chasek, "Power Politics, Diplomacy and Role Playing: Simulating the UN Security Council's Response to Terrorism," *International Studies Perspectives* 6/1 (2005): 1–19.

30. J. Lobasz and B.Valeriano, "Teaching (International) Politics with Science Fiction Literature and Films," in *Handbook of Teaching and Learning in Political Science and International Relations*, J. Ishiyama, W.J. Miller, and E. Simon, eds. (Edward Elgar: Cheltenham, 2015), 399–410.

WORKS CITED

American Political Science Association. 1951. *Goals for Political Science: Report of the Committee for the Advancement of Teaching, American Political Science Association.* New York: William Sloane Associates.

Baumann, Z.D., K. Marchetti, & B. Soltoff. 2015. "What's the Payoff?: Assessing the Efficacy of Student Response Systems." *Journal of Political Science Education* 11/3: 249–63.

Bernstein, J. 2012. "Signature Pedagogies in Political Science: Teaching Students How Political Actors Behave." In *Exploring More Signature Pedagogies: Approaches to Teaching Disciplinary Habits of Mind*, N.L. Chick, A. Haynie, & R.A.R. Gurung, eds. Sterling, VA: Stylus.

Biggs, J. & C. Tang. 2007. *Teaching for Quality Learning at University*. Maidenhead: McGraw Hill.

Blight, D.A. 2012. *What's the Use of Lectures?* Harmondsworth: Penguin.

Brown, S. & P. Race. 2002. *Lecturing a Practical Guide*. London: Kogan Page.

Chasek, P.S. 2005. "Power Politics, Diplomacy and Role Playing: Simulating the UN Security Council's Response to Terrorism." *International Studies Perspectives* 6/1: 1–19.

Craig, J. 2012. "What (if Anything) is Different about Teaching and Learning in Politics?" In *Teaching Politics and International Relations*, C. Gormley-Heenan & S. Lightfoot, eds. Basingstoke: Palgrave Macmillan: 22–37.

Dougherty, B.K. 2003. "Byzantine Politics: Using Simulations to Make Sense of the Middle East." *PS: Political Science and Politics* 36/2: 239–44.

Dunleavy, P. 1986. *Studying for a Degree in the Humanities and Social Sciences.* Basingstoke: Macmillan.

Easton, D., J.G. Gunnell, and M.B. Stein, eds. 2017. *Regime and Discipline: Democracy and the development of political science.* Ann Arbor: University of Michigan Press.

Farr, J. & R. Seidelman, eds. 1993. *Discipline and History: Political Science in the United States.* Ann Arbor: University of Michigan Press.

Glover, R.W. 2013. "Introduction: 'Like Captured Fireflies…'" in *Teaching Politics Beyond the Classroom*, R. Glover & D Tagliarina, eds. New York: Bloomsbury: 1–6.

Harris, C. & B. Quinn, 2015. "Teaching, Assessment and Professional development: Praxis in Ireland's Political Science Community." *Irish Political Studies* 32/2: 255–75.

Hartlaub, S.G. & F.A. Lancaster. 2008. "Teacher Characteristics and Pedagogy in Political Science." *Journal of Political Science Education* 4/4: 377–93.

Hassner, R.E. 2005. "Sliding into Home Plate: How to Use Slideware to Improve Your Presentation (While Dodging the Bullets)." *PS: Political Science and Politics* 38/3: 393–97.

Holland, L., P. Schwartz-Shea, & Jennifer M.J. Yim. (2013). "Adapting Clicker Technology to Diversity Courses: New Research Insights." *Journal of Political Science Education* 9/3: 273–91.

Huerta, J.C. 2007. "Getting Active in the Large Lecture." *Journal of Political Science Education* 3/3: 237–49.

Jenkins, S. 2015. "Flipping the Introductory American Politics Class: Student Perceptions of the Flipped Classroom." *PS: Political Science and Politics* 48/4: 607–11.

Krain, M. & J.S. Lantis. 2006. "Building Knowledge? Evaluating the Effectiveness of the Global Problems Summit Simulation." *International Studies Perspectives* 7/4: 395–407.

Laurillard, D. 2002. *Rethinking University Teaching: A Conversational Framework for the Effective Use of Learning Technologies*. New York: Routledge Falmer.

Laves, W. 1940. "The Next Decade in Political Science Teaching." *American Political Science Review* 34/95: 983–6.

Leftwich, A. 1987. "Room for Manoeuvre: A Report on Experiments in Alternative Teaching and Learning Methods in Politics." *Studies in Higher Education* 12/3: 311–23.

Leston-Bandeira, C. 2012 "Enhancing Politics Teaching through Active Learning." In *Teaching Politics and International Relations*, Cathy Gormley-Heenan & Simon Lightfoot eds. Basingstoke: Palgrave Macmillan: 51–64.

Lobasz, J. & Valeriano, B. 2015 "Teaching (International) Politics with Science Fiction Literature and Films." In *Handbook of Teaching and Learning in Political Science and International Relations*, J.T. Ishiyama, W.J. Miller, & E. Simon, eds. Edward Elgar: Cheltenham: 399–410.

Mallinson, D. & Z. Bauman. 2015. "Lights, Camera, Learn: Understanding The Role Of Lecture Capture In Undergraduate Education." *PS: Political Science and Politics* 48/3: 478–82.

Mayhew, E. 2018. "The New Generation of Prezi Presentation Software, Provided by Prezi Inc., and Student Engagement and Learning Within Political Science." *Journal of Political Science Education*. Published Online: November 8, 2018, DOI: https://doi.org/10.1080/15512169.2018.1485572.

McKechnie, W.J. 1999. *McKechnie's Teaching Tips: Strategies, Research, and Theory for College and University Teachers*. Boston: Houghton Mifflin Company.

Moore, M.J. 2011 "How (and What) do Political Theorists Teach: Results of a National Survey." *Journal of Political Science Education* 7/1: 95–128.

Omelicheva, M.Y. & O. Avdeyeva. 2008. "Teaching with Lecture or Debate? Testing the Effectiveness of Traditional Versus Active Learning Methods of Instruction." *PS: Political Science and Politics* 41/3: 603–7.

Roberts, J.C. 2015. "Evaluating the Effectiveness of Lecture Capture: Lessons Learned from an Undergraduate Political Research Class." *Journal of Political Science Education* 11/1: 45–60.

Robson, W.A. 1951. *The University Teaching of Social Sciences: Political Science*. Paris: UNESCO.

Schrad, M.L. 2010 "In Defense of the Populist Lecture." *PS: Political Science and Politics* 43/4: 759–65.

Stammers, N., H. Dittmar, & J. Henney. 1999. "Teaching and Learning Politics: A Survey of Practices and Change in UK Universities." *Political Studies* 47/1: 114–26.

Touchton, M. 2015. "Flipping the Classroom and Student Performance in Advanced Statistics: Evidence from a Quasi-Experiment." *Journal of Political Science Education* 11/1: 28–44.

Velasco, M. & G. Çavdar 2013. "Teaching Large Classes with Clickers: Results from a Teaching Experiment in Comparative Politics." *PS: Political Science & Politics* 46/4: 823–29.

Woodcock, P. & G. Duckworth. 2010. "iPod Therefore I Am: Using PC Videos to Aid the Teaching of the History of Political Philosophy." *European Political Science* 9/1: 25–33.

Chapter 11

The Lecture in Political Science

Lee Trepanier

In a 1951 report about teaching political science, the American Political Science Association called to revitalize and modernize teaching political science, including the lecture which had appeared to have outlived its usefulness.[1] However, thirty-five years later, Patrick Dunleavy provided a defense of the lecture because of its accessibility, flexibility, and ability to demonstrate to students how an argument can be constructed in real time.[2] More recently, the discipline has seen new methods of instruction, whether technology or active learning techniques, adopted by those who lecture with varying degrees of success in student learning.[3] The popularity of the online lectures about justice by Michael Sandel and others also seems to confirm the sustaining power of the lecture as an effective pedagogical tool.[4] Finally, a 2015 study that compared students enrolled in a lecture-only undergraduate research course with those enrolled online revealed the former scored significantly higher in both grades and course completion.[5] Put succinctly, the lecture has been modernized and continues to remain an essential component in teaching students political science.

However, what has not been explored in the discipline is whether the lecture, even with the adoption of technology and active learning, is an effective way to teach students according to the subfields of American politics, comparative politics, international relations, and political philosophy.[6] Each of these subfields pose unique challenges to those who lecture and therefore require different ways to overcome them in teaching students. In this chapter, I will address these issues. While not pretending that I will identify every relevant issue for each subfield, I do hope that this chapter will prompt more thinking and research about how political scientists lecture according to subfield.

WHAT IS A LECTURE?

But first what is a lecture? The past participle of the Latin word *legere* is *lectum* which means to collect or gather things together. Initially it was used to gather food, like grapes, but later meant to select judges, and then, to read or peruse books, reading aloud what the lecturer had chosen for a specific occasion.[7] Today the lecture contains four elements: the lecturer, the material, the audience, and the setting. To lecture effectively is to be aware of these four factors, such as presenting material that is intended to be read aloud to an audience rather than material that is intended only to be read or adjusting the content of the lecture depending whether the audience is undergraduate students or one's academic colleagues.

The setting of the lecture also plays a pivotal role in how to lecture: a classroom of 30 students can make the lecture more conversational than a lecture hall of 300, lending the form to a more formal presentation. Video recording of lectures is an instance where the lecture is no longer performed in real time but can be rehearsed until the lecturer is satisfied with the result. The adoption of technology for the lecture, like visual aids, depends upon whether they are available to the lecturer, for example, traditional versus smart classrooms. In many ways, the lecture is shaped by its setting as much as it is by its audience and material.

Finally, the personality and the character of the lecture counts in the lecture's effectiveness: does the speaker have authenticity and authority when lecturing or come across as smug, insincere, or clueless? In Aristotle's *Rhetoric*, there are three types of *ethos* that are described when speaking: (1) *phronesis* (practical wisdom), (2) *arete* (virtue), and (3) *eunoia* (goodwill toward the audience).[8] These three *ethe* can be translated today as knowledge, authenticity and authority, and likeability. While there are other *ethe* than Aristotle's, these three at least provide a baseline of how lecturers should think about presenting their material in public.

CHALLENGES TO THE POLITICAL SCIENCE LECTURE

There are several challenges to the lecture which have been mentioned in this book: it promotes passive learning, it requires professors to have effective speaking skills, it is an outdated way to convey information in the digital age, it has transformed itself today to entertain rather than to instruct, it is not an effective method of student learning or a way to recruit students into one's program. In addition to these problems, the lecture confronts the obstacles specific to political science: What is the relevance of the material being presented, what is true and not true in this era of "fake news," how to present material when students lack historical and comparative political knowledge, and what is the larger context of what is being learned?

These challenges manifest themselves uniquely in each subfield of political science. For the lecture in international relations and comparative politics, students often lack basic historical and comparative knowledge, thereby making the lecture more historical instead of analytical in nature. For the lecture in American politics, students not only confuse political knowledge for opinion but also speculate about the political bias of the lecturer which further negates any information being conveyed. And for political philosophy, students fail to see the relevancy to their own lives and politics of being taught the theories and texts of Plato, Aquinas, Marx, and others.[9]

Below I have created a table to summarize these issues (table 11.1).

Table 11.1 The Political Science Lecture Subfields

Subfield	*Background Knowledge*	*Content Knowledge*	*Theoretical Knowledge*	*Trustworthiness*	*Relevance*
International Relations	Yes	Yes	Yes	No	No
Comparative Politics	Yes	Yes	Yes	No	No
American Politics	Yes	Yes	Yes	Yes	No
Political Theory	No	No	No	No	Yes

Background knowledge (historical and comparative information, for example, the history of nineteenth-century Japan) and content knowledge (information the actual subject to be studied, for example, the workings of Congress) are problems particular to the lecture in international relations, comparative politics, and American politics.[10] Students lack the necessary knowledge about history or other societies and therefore time is spent on these subjects before the actual study of these topics can commence. This can be an issue for political philosophy for those who adopt the Cambridge school approach, which looks at the historical context of political theories and texts. But, for most political philosophers, they focus on the text itself, thereby making background knowledge less of an obstacle to overcome and leaving ample time to lecture about the content itself.

Without background and content knowledge, theoretical knowledge (how background and content knowledge are connected to the theories of the subject, for example, how does the international relations theory of realism explain the outbreak of World War I) is impossible to lecture in international relations, comparative politics, and American politics.[11] Lecturers may have to spend most of their semester providing the necessary background and content knowledge before moving to theoretical knowledge. Again, this is less of an issue for the textual analysis approach in political philosophy where minimal background knowledge is needed and therefore more time can be spent on content and theoretical knowledge, for example, how Voegelin interprets Plato's *Republic* compared to Strauss.

The other two areas that I mention are trustworthiness of the claims that are made in the lecture, whether they are in background, content, and theoretical knowledge, and the relevancy of the topic to students. With the rise of "fake news" and the availability of information via the Internet, students can quickly find information that challenges what the lecturer claims (whether the information on the Internet is accurate is often immaterial to students). This is particularly true for those who lecture about American politics given today's partisan climate of opinion. This is less of a problem, if at all, for those who lecture about international relations or comparative politics since the students' basic knowledge about these subjects is minimal.

This also is not a problem for political philosophy, since students are interested in politics because of "current events" rather than "dead texts.[12] I have yet to come across an undergraduate student who questions my understanding of Gadamer in reading Plato (although I certainly would welcome it!). However, the lack of the perceived "currentness" in political philosophy texts makes it difficult to show their relevancy to students, whereas this is not a problem for those who lecture in international relations, comparative politics, and American politics.

Granted this typology of the challenges to the lecture in the political science subfields is incomplete. It is also aimed for undergraduate students, since graduate students are usually taught in a seminar rather than lecture format, and to the general public instead of academic colleagues, where background, content, and theoretical knowledge is not an issue for them. Nevertheless, I hope this chart paves a path for us to think about the key obstacles to the lecture—and teaching more generally—in the political science subfields and the ways we can surmount them.

SUGGESTIONS FOR THE POLITICAL SCIENCE LECTURE

One way to think about how to make the lecture more effective in student learning is to ask oneself, in the spirit of backward course design, what are the outcomes one wants out of the lecture, what lecture techniques can best accomplish this, and at what knowledge level are the students entering the classroom.[13] Although the goals may vary for each subfield, the lecture in political science generally should impart knowledge (background, content, and theoretical), skill sets (the ability to comprehend, analyze, and synthesize information), and an interest in politics (civic education).[14] With these three objectives—knowledge, skills, and interest—the professor can determine which lecture techniques will best help achieve these goals.

One technique to keep in mind is Blight's recommendations: (1) students' memory can only retain on average seven items; (2) key points should be made in less than two seconds to avoid students forgetting; and (3) silence is critical so as not to create inference between what was said and what comes next.[15] Blight also mentions that the professor should not assume that his or her own experiences are the same as that of students and thus the professor must find new ways to create a rapport with them. Lastly, the professor should organize the lecture so it is accessible, organized, and concise in its presentation and signal to students when important points are to be mentioned.

Nilson suggests the incorporation of visual examples in the lecture to showcase key findings.[16] She also calls for professors to incorporate a variety of active learning techniques throughout the lecture: pairing activities where students compare lecture notes and can ask questions; reflection paragraph assignments; teamwork exercises to solve a problem or discuss a subject; multiple test quizzes; simulation activities; and the use of flashcards, hands, and clickers for student input. Other techniques involve the use of question-based outlines, student surveys, and discussion-prompt questions; recording and evaluating one's own lecture before delivering it; and the use of certain technologies like PowerPoint and Keynote.[17]

Davis suggests observation of other excellent lecturers and learn how not to read your lecture when delivering it.[18] When opening a lecture, signal to students that the lecture is beginning and grab their attention with an interesting or dramatic opening. After this is established, you want to continue to announce your objectives for the class and the organization of the lecture. While lecturing itself, you want to vary your delivery to keep students' interest; be enthusiastic, conversational, and use examples to underscore pivotal points. You also want to be conscious of your own presentation—project your voice, maintain eye contact with students, and use movement to hold their attention—while keeping track of the time so you don't have to rush any part of your lecture. Finally, you want to close your lecture forcefully with a conclusion and, afterward, make notes about your own performance.

All of these techniques can be adapted by professors to enhance the effectiveness of their political science lectures. But for the specific subfields of international relations, comparative politics, and American politics, I also would suggest that the problems of background and content knowledge can be partially addressed by requiring online quizzes or assignments before students meet in class, thereby forcing them to be familiar with the material. It also allows professors to assess how much students have understood before the lecture. For example, students in an American Politics course are required to take online quizzes before a class about the history and evolution of the

U.S. Constitution. The professor could see not only the students who have read the material but what areas students did not understand and thus adjust the lecture accordingly. This would give the professor more time to devote to content and theoretical knowledge rather than background knowledge.

For the political philosophy lecture, the problem of relevance can be addressed by showing how certain theories, concepts, and ideas derived from texts can be applied and evaluated in the contemporary world. This makes the use of examples and stories as well as opening and closing the lecture with a dramatic account critical for a political philosophy lecture to be effective in student learning. Evaluating the merits and deficiencies of certain theories and ideas applied to current events is another useful exercise. Professors must demonstrate the application and evaluation of political philosophy to contemporary politics to make their lectures relevant. For the subfields of international relations, comparative politics, and American politics, the application and evaluation of knowledge is already explicit in their lectures, for example, to study the U.S. Supreme Court is to understand abortion rights, to learn political economy is to know a country's trade policy. However, in a political philosophy lecture, the relevance of the text is implicit and therefore the professor must make it explicit to the student.

Finally, with regard to truth claims, professors must look to the *ethos* of themselves as lecturers and the audience. They must show their practical wisdom, virtue, and goodwill toward students in the lecture, demonstrating their expertise and possessing a directed interest in the lecture rather than an ulterior motive. They want to avoid the rehearsed smugness and weird guru presence of TED and YouTube speakers and instead opt for humility, good-naturedness, and, when appropriate, humor. Finally, professors probably should not reveal their politics in the lecture and present the material even-handily, thereby mitigating potential accusations about political bias from students. All these techniques can help make the claims made in the lecture credible to students.[19]

CONCLUSION

When we think about the lecture, we need to take into account the lecturer, the material, the audience, and the setting. With regard to political science, we need to think about the lecture in specific subfields, with each subfield posing unique challenges to the lecture. How the professor overcomes these obstacles—whether they are in background, content, and theoretical knowledge; trustworthiness of the claims made; or relevance of the topic—requires different strategies for each subfield. In the hope of continuing this conversation, this chapter suggests some possible strategies and techniques to make the lecture more effective in each subfield.

Having done that, it is important to keep in mind that at its fundamental core the lecture is a type of trust between the professor and the student. It is not a debate, where one side attempts to emerge victorious at the expense of another or a conversation which includes the utilitarian and trivial. The lecture differs from the seminar where multiple views are exchanged and the Socratic Method where students are questioned, whether individually or as a group, by someone who does not offer knowledge but asks for it.[20] The lecture is a type of forum that invites people to listen collectively about the highest and most important things in life with no winners or losers after it is finished. It presents a singular perspective but prompts others to devise their own afterward, avoiding the confusion and cacophony of the seminar and the Socratic Method. In many ways, a lecture is like a prayer, an invocation that seeks a rapport with something divine through deliberate communication with others; and, like all prayers, it is rooted in the hope that the truth can be shared among all, for only the gods own wisdom.

NOTES

1. American Political Science Association, *Goals for Political Science: Report of the Committee for the Advancement of Teaching, American Political Science Association* (New York: William Sloane Associates, 1951).

2. Patrick Dunleavy, *Studying for a Degree in the Humanities and Social Sciences* (Basingstoke: Macmillan 1986); also see John Craig, "What (if Anything) is Different about Teaching and Learning in Politics?" in *Teaching Politics and International Relations*, C. Gormley-Heenan and S. Lightfoot, eds. (Basingstoke: Palgrave Macmillan, 2012), 31–34 and his chapter in this volume for more about the history of the lecture in the discipline of political science; also see Lee Trepanier, "SoTL as a Subfield for Political Science Graduate Programs," *Journal of Political Science Education* 13/2 (2017): 138–51.

3. For technology and the lecture, see Mark Lawrence Schrad, "In Defense of the Populist Literature," *PS: Political Science & Politics* 43/4 (2010): 759–65; Marcela Velasco and Gamze Çavdar, "Teaching Large Classes with Clickers: Results From a Teaching Experiment in Comparative Politics," *PS: Political Science and Politics* 46/4 (2013): 823–29; Daniel J. Mallinson and Zachary D. Baumann, "Lights, Camera, Learn: Understanding the Role of Lecture Capture in Undergraduate Education," *PS: Political Science & Politics* 48/3 (2015): 478–82; Zachery D. Baumann, Kathleen Marchetti, and Benjamin Soltoff, "What's the Payoff?: Assessing the Efficacy of Student Response Systems," *Journal of Political Science Education* 11/3 (2015): 249–63; Heather K. Evans and Victoria Cordova, "Lecture Videos in Online Courses: A Follow-Up," *Journal of Political Science Education* 11/4 (2015): 472–82.

For active learning and the lecture, see Juan Carlos Huerta, "Getting Active in the Large Lecture," *Journal of Political Science Education* 3/3 (2007): 237–49; Robert P. Amyot, "Is It Better to Be Feared Than Loved? Investigating the Efficacy of Different Teaching Methods on the Learning of Specific Content." *PS: Political Science & Politics* 47/4 (2014): 855–61.

4. Michael Sandel, "What's the Right Thing to Do?" *Harvard University's Justice with Michael Sandel*. Available at http://justiceharvard.org/; also see Thomas L. Friedman, "Justice Goes Global," *New York Times*, June 4, 2011. Available at https://www.nytimes.com/2011/06/15/opinion/15friedman.html.

5. James C. Roberts, "Evaluating the Effectiveness of Lecture Capture: Lessons Learned from an Undergraduate Political Science Research Class," *Journal of Political Science Education* 11/1 (2015): 45–60.

6. This list is the traditional four subfields of political science, although programs vary in their offering with additions like public law, public policy, methods, and other areas of concentration.

7. James V. Schall, "What is a Lecture?" in *Another Sort of Learning: Selected Contrary Essays on the Completion of Our Knowledge or How Finally to Acquire an Education While Still in College, or Anywhere Else: Containing Some Belated Advice about How to Employ Your Leisure Time When Ultimate Questions Remain Perplexing in Spite of Your Highest Earned Academic Degree, Together with Sundry Book Lists Nowhere Else in captivity to Be Found* (San Francisco: Ignatius Press, 1978), 188–89.

8. Aristotle, *Rhetoric* (Oxford: Oxford University Press, 2002). Aristotle also points out that the audience has a type of *ethos* too, of which the speaker must be aware: (1) the speaker has a direct interest in the speech or outcome of the debate; (2) the speaker has a vested interest or ulterior motive; and (3) the speaker possesses or does not possess expertise of the subject.

For more about prudence, see Lee Trepanier, "A Philosophy of Prudence and the Purpose of Higher Education Today," in *The Relevance of Higher Education*, Timothy Simpson, ed. (Lexington Books, 2013), 1–23 and "The Character Model for the American University," *Expositions* 12/2 (2018): 132–51.

9. For more about the history and challenges confronting political philosophy, see Lee Trepanier, "The Relevance of Political Philosophy and Political Science," in *Why the Humanities Matter Today: In Defense of Liberal Education* (Lexington Books, 2017), 127–44.

10. Another way to think about background and content knowledge is the comprehension of the subject (background knowledge) and the analysis of knowledge (content knowledge).

11. One way to think about theoretical knowledge is the synthesis of knowledge which requires background (comprehension) and content (analysis) knowledge.

12. Jonas B. Bunte, "Why Do Students Enroll in Political Science Courses?" *PS: Political Science & Politics*, January 29, 2019. Available at https://politicalsciencenow.com/why-do-students-enroll-in-political-science-courses/.

13. Grant Wiggins and Jay McTighe, *Understanding by Design* (Alexandria, VA: Association for Supervision & Curriculum Development, 2005).

14. For more about civic education, see Lee Trepanier, "Aristotelian Pluralism and Diversity: The Conditions of Civic Education and the Common Good," *Expositions* 8/1 (2014): 100–21 and "Aristotelian Citizenship for a Multicultural World," in *Citizenship and Multiculturalism in Western Liberal Democracies*, David Edward Tabachnick and Leah Bradshaw, eds. (Lexington Books, 2017), 1–21.

15. Donald Blight, *What's the Use of the Lecture?* (Harmondsworth: Penguin, 1972).

16. Linda B. Nilson, *Teaching at Its Best: A Research-Based Resource for College Instruction* (San Francisco: Jossey-Bass, 2010), 113–25.

17. Huerta, "Getting Active in the Large Lecture"; Evans and Cordova, "Lecture Videos in Online Courses"; Mallinson and Baurmann, "Lights, Camera, Learn: Understanding the Role of the Lecture Capture in Undergraduate Education"; Schard, "In Defense of the Populist Lecture."

18. Barbara Gross Davis, *Tools for Teaching* (San Francisco: Jossey-Bass, 2009), 148–60.

19. For an example of political science professors who accomplish this, see John von Heyking and Lee Trepanier, eds. *Teaching in an Age of Ideology* (Lanham: Lexington Books, 2012).

20. For more about the Socratic Method, see Lee Trepanier, ed. *The Socratic Method Today: Student-Centered and Transformative Teaching in Political Science* (New York: Routledge, 2017).

WORKS CITED

American Political Science Association. 1951. *Goals for Political Science: Report of the Committee for the Advancement of Teaching, American Political Science Association*. New York: William Sloane Associates.

Amyot, Robert P. "Is it Better to Be Feared Than Loved? Investigating the Efficacy of Different Teaching Methods on the Learning of Specific Content." *PS: Political Science & Politics* 47/4 (2014): 855–61.

Aristotle. 2002. *Rhetoric*. Oxford: Oxford University Press.

Baumann, Zachery D., Kathleen Marchetti, and Benjamin Soltoff. 2015. "What's the Payoff?: Assessing the Efficacy of Student Response Systems." *Journal of Political Science Education* 11/3: 249–63.

Blight, Donald. 1972. *What's the Use of the Lecture?* Harmondsworth: Penguin.

Bunte, Jonas B. 2019. "Why Do Students Enroll in Political Science Courses?" *PS: Political Science & Politics*, January 29. Available at https://politicalsciencenow.com/why-do-students-enroll-in-political-science-courses/.

Craig, John. 2012. "What (if Anything) is Different about Teaching and Learning in Politics?" In *Teaching Politics and International Relations*, C. Gormley-Heenan and S. Lightfoot, eds. Basingstoke: Palgrave Macmillan: 31–34.

Davis, Barbara Gross. 2009. *Tools for Teaching*. San Francisco: Jossey-Bass.

Dunleavy, Patrick. 1986. *Studying for a Degree in the Humanities and Social Sciences*. Basingstoke: Macmillan.

Evans, Heather K. and Victoria Cordova. 2015. "Lecture Videos in Online Courses: A Follow-Up." *Journal of Political Science Education* 11/4: 472–82.

Friedman, Thomas L. 2011. "Justice Goes Global." *New York Times*, June 4. Available at https://www.nytimes.com/2011/06/15/opinion/15friedman.html.

Heyking, John von and Lee Trepanier, eds. 2012. *Teaching in an Age of Ideology*. Lanham: Lexington Books.

Huerta, Juan Carlos. 2007. "Getting Active in the Large Lecture." *Journal of Political Science Education* 3/3: 237–49.

Mallinson, Daniel J. and Zachary D. Baumann. 2015. "Lights, Camera, Learn: Understanding the Role of Lecture Capture in Undergraduate Education." *PS: Political Science & Politics* 48/3: 478–82.

Nilson, Linda B. 2010. *Teaching at Its Best: A Research-Based Resource for College Instruction*. San Francisco: Jossey-Bass.

Sandel, Michael. n.d. "What's the Right Thing to Do?" *Harvard University's Justice with Michael Sandel*. Available at http://justiceharvard.org/.

Schall, James V. 1978. "What is a Lecture?" In *Another Sort of Learning: Selected Contrary Essays on the Completion of Our Knowledge or How Finally to Acquire an Education While Still in College, or Anywhere Else: Containing Some Belated Advice about How to Employ Your Leisure Time When Ultimate Questions Remain Perplexing in Spite of Your Highest Earned Academic Degree, Together with Sundry Book Lists Nowhere Else in captivity to Be Found*. San Francisco: Ignatius Press: 188–89.

Schrad, Mark Lawrence. 2010. "In Defense of the Populist Literature." *PS: Political Science & Politics* 43/4: 759–65.

Trepanier, Lee. 2013. "A Philosophy of Prudence and the Purpose of Higher Education Today." In *The Relevance of Higher Education*, Timothy Simpson, ed. Lexington Books, 2013: 1–23.

———. "Aristotelian Pluralism and Diversity: The Conditions of Civic Education and the Common Good." *Expositions* 8/1 (2014): 100–21.

———. 2017. "Aristotelian Citizenship for a Multicultural World." In *Citizenship and Multiculturalism in Western Liberal Democracies*, David Edward Tabachnick and Leah Bradshaw, eds. Lanham: Lexington Books: 1–21.

———. 2017. "SoTL as a Subfield for Political Science Graduate Programs." *Journal of Political Science Education* 13/2: 138–51.

———. "The Character Model for the American University." *Expositions* 12/2 (2018): 132–51.

———. 2017. "The Relevance of Political Philosophy and Political Science." In *Why the Humanities Matter Today: In Defense of Liberal Education*, Lee Trepanier, ed. Lexington Books: 127–44.

———. ed. 2017. *The Socratic Method Today: Student-Centered and Transformative Teaching in Political Science*. New York: Routledge.

Velasco, Marcela and Gamze Çavdar. 2013. "Teaching Large Classes with Clickers: Results From a Teaching Experiment in Comparative Politics." *PS: Political Science and Politics* 46/4: 823–29.

Wiggins, Grant and Jay McTighe. 2005. *Understanding by Design*. Alexandria, VA: Association for Supervision & Curriculum Development.

Chapter 12

The Music Lecture as Gestalt

Jamie Fiste

Psychologists have long recognized that student learning is facilitated when the details and subdivisions of any topic are understood in relation to larger contexts, connections are made within the "parts and the "whole," and patterns are recognized within a domain of study. Students, however, in the beginning stages of learning, lack the prerequisite knowledge to construct larger frames of reference important for learning. Even more advanced students benefit greatly from a professor who is able to draw connections in inspiring ways. These points have tremendous implications for the value of the music lecture in the hands of a professor who expertly "placards" a larger picture or "gestalt" for the students and makes connections between seemingly isolated events. I hope to persuade the reader that in all levels of learning, the lecture can provide context and a motivational example of organized thought, thereby enhancing learning, critical thinking, and memory performance.

Learning can be defined as a change in long-term memory[1] and a relatively permanent change in behavior that can be objectively observed.[2] This observable behavior could be anything from the avoidance of a hot surface, taking an exam, or playing a violin. Any discussion regarding the effectiveness of the lecture needs to address the question of whether it is an effective tool to effect long-term change in the learner. If not, then other tools must be sought.

Since learning is defined as a change in long-term memory (LTM), it is worthwhile to examine studies regarding the nature of LTM. Through an array of creative experiments, scientists have generally concluded that the brain stores information in the form of *schemas*: knowledge structures that organize information in a hierarchical and interconnected way.[3] Piaget referred to schema as a "construction of reality" that is used to interpret new events.[4] The inherent organization and interconnectedness of knowledge in

LTM has strong implications for learning. For instance, by the end of the 1970s, it was already observed that people are more likely to remember information if it is connected to a central theme than information that is not.[5]

Going back to the 1930s, Frederick Bartlett (1886–1969) who coined the term "schema," concluded that attaching meaning to information plays a fundamental role in memory. He referred to this process as an *effort after meaning*.[6] In other words, memory is selective—people tend to remember that which seems most important to them—and interpretive—people attach meaning to what they remember.[7] Interpretation and meaning can be as simple as assuming a bad motive for being cut off in traffic to a more sophisticated perspective of viewing music from the late sixteenth century to the late twentieth century through the lens of modernity.[8]

Gestalt psychology, which began in the early twentieth century, observed that people tend to perceptually organize their environment into patterns or by grouping "parts" into "wholes." Perhaps you have noticed this tendency in yourself when you saw that rabbit or dog in the clouds. Experiments continued and in 1941 it was observed that young students, when presented with new material, organized and simplified it into meaningful patterns.[9] Similarly, Cronbach noted in the 1950s that people look for the simplest patterns to summarize main facts.[10]

Gestalt psychologist Max Wertheimer wrote, "In human terms there is at the bottom the desire, the craving to face the true issue, the structural core, the radix of the situation."[11] Some will immediately discern the humanistic thrust of Wertheimer's statement. In fact, Gestalt psychology was the first humanistic psychology, unpacking perception from a distinctly human perspective. Part of being human, in virtual all developmental levels, is to make order from chaos, construct patterns, affix meaning, make associations, and establish connections between events and concepts.

Max Wertheimer (1880–1943), Kurt Koffka (1886–1941), and Wolfgang Köhler (1887–1967) were credited with this new approach to psychology that contrasted with the dominant behaviorist model at the time. Gestalt is roughly translated from the German as "form," "configuration," or "pattern" and may be better known by Kurt Koffka's famous summary, "The whole is different from the sum of its parts." The more popular rendition of "The whole is greater than the sum of its parts" is an incorrect translation of Koffka's phrase and has resulted in confusion.[12] "Greater than" may imply a value judgment, which is not the intention at all. "Different from" simply means that the attributes of the whole are distinct from the attributes of the individual parts.

To get at what Koffka meant, music of the common practice period (ca. 1650–1900) can provide an illustration. One way to understand music of this era is as a complex set of interactions between rhythm, harmony, timbre, texture, form, melody—to name a few—that result in a new set of attributes that comprise the whole. From this perspective, all the individual parts contribute to, and can be said to, derive "meaning" from the whole. These parts can be further subdivided into "smaller wholes" with their own inherent interactions. For instance, the resulting sound of combining instruments together or "orchestration" is distinct from the sounds of the individual instruments in isolation. Similarly, a melody has its own characteristics that differentiate it from the isolated notes and rhythms that comprise it: shape, intervals, and pacing. One can further subdivide these melodic parts into smaller wholes. Pitches, for example, are comprised of fundamentals and overtones that together formulate the tone color we hear in conjunction with the pitch. Moreover, rhythms unfold and must be understood by their relationship to time.

To take this idea further, individual compositions, while wholes unto themselves, are part of still larger contexts. Works of individual composers can be grouped into compositional periods, such as Mahler's folk, neoclassical, and eclectic periods. To widen the lens even further, composers' outputs are part of larger stylistic eras, such as the romantic, classical, and baroque periods. Stylistic periods also find their context in economic trends, such as the growth of the middle class during the eighteenth century, intellectual movements such as sixteenth-century humanism, and catastrophic events such as World War II, all of which were deeply felt within the arts. Replete throughout history are complex interactions between extra musical events and compositional styles. These interactions are part of sophisticated knowledge structures expressed in differing musicological theories and interpretations. This is human nature on display: the craving and searching for more accurate and complete organizations of complex phenomena. The expert eventually weaves an astonishingly complex web of hierarchies, wholes, parts, relations, associations, and interactions that comprise an impressive gestalt stored in the schema of LTM.

The highly organized schema in experts has several benefits relating to memory, one being superior recall of information related to their field. For instance, experts often have better recall of episodic information (i.e., personal experiences) related to their domain of study:

> Routine experts differ from novices both in the structure of domain-related knowledge and in the ability to retrieve episodic domain-related information . . . . With certain exceptions, experts exhibit superior episodic memory for domain-typical information largely because they can directly access LTM to rapidly and reliably encode and retrieve the information, rather than maintaining it in short term memory. (STM)[13]

Episodic memory is not the only area where recall is improved in experts. In a study done by Chase and Simon (1973), chess masters were better able to remember chess pieces in middle and end game positions after a five-second exposure time than novices: 90 percent accuracy verses 40 percent.[14] The advantage disappeared if the chess pieces were not organized according to patterns seen in chess games (i.e., "domain-typical" patterns). The improved memory specific to domain-related patterns is not only regulated to chess but can be found in many other fields as well.[15] One researcher summarized the findings of the study:

> Examined in more detail, the findings suggested that the advantage enjoyed by the chess masters stemmed from their ability to perceive the chessboard as an organized whole, rather than as a collection of individual pieces.[16]

So complex schemas aid the expert in recall and learning. This is all wonderful for the expert but most of our students are not yet at this level. The good news is that when new information is organized at the time of learning—in a sense providing an external schema—novice learners display better recall. In a study done by Bransford and Johnson, participants were given this passage without any pre-organizing information:[17]

> The procedure is actually quite simple. First you arrange items into different groups. Of course one pile may be sufficient depending on how much there is to do. If you have to go somewhere else due to the lack of facilities that is the next step; otherwise you are pretty well set. It is important not to overdo things. That is, it is better to do too few things at once than too many.[18]

Not surprisingly, this passage was difficult for the participants to comprehend and recall even when the title, *Washing Clothes*, was offered after the fact. Only when the title was given in advance was comprehension and recall improved.[19]

To summarize, existing evidence shows that the human brain stores information in LTM, not as isolated and random bits of information, but in hierarchically organized schemas that are interconnected. We tend to remember new information if we can relate it to an existing schema or are provided with an organizational framework at the time of learning. Memory and therefore learning is intimately tied to hierarchical organization, organized wholes, relating facts to a central theme, context, patterns, interpretation, interconnectedness, meaning, and associations.

Having provided a brief background on memory and learning, the question remaining is how the instructor of music history or theory can make applications to teaching. Considering what has been presented thus far, one way to view education is the *assisting of students in organizing their thoughts*. This applies to novice students or those pursuing advanced degrees. The strategies for this endeavor can be subdivided into two main parts:

1. Providing students with existing structures of thought in a domain of study;
2. Challenging students to analyze and question these structures to improve upon them or to come up with theories of their own.

The first of the two proposals, providing students with existing structures of knowledge, can be done through readings, videos, and lectures by the professor or guest speakers. As mentioned previously, the professor can provide for novice students what they are unable to give themselves: a context for new information. Moreover, advanced students, even professors, can be inspired and motivated by an expert presentation that offers a fresh perspective on a domain of study. I imagine many reading this book have experienced exactly what I am referring to. Many of us admire scholars in our field that have made tremendous impacts on our intellectual life. We are drawn to their clarity of thought, perceptiveness, and architecture of their argument. They are great examples to motivate our own intellectual pursuits.

A well-organized system of thought has its own innate beauty that sparks the imagination much like a great work of art or music. The association between speech and music is well documented, not just because music of differing cultures seems to reflect the characteristics of speech from those respective countries, but because many composers and orators throughout history have seen a close connection between music and the spoken arts. This connection has origins in the ancients and is evident during the baroque period when even instrumental music was understood as an expression of rhetorical techniques and aims. Rhetoric and music are close cousins with overlapping purposes. Music and rhetoric alike make a case, seek to persuade, and "move the affections." Could not the lecture at minimum be a source of motivation and inspiration for students? Perhaps the growing lack of interest in the lecture has its roots in the lost art of oratory that was once standard for all education. An outgoing and flamboyant personality is not necessary for a well-delivered presentation; the power of ideas is itself compelling, evidenced by many popular TED Talk lectures.

The second proposal, that education involves encouraging students to analyze and question existing intellectual structures or theories, could be the domain of what is today called "active learning." Max Wertheimer, in his book *Productive Thinking,* recounts hours spent with Einstein to determine the steps in his thinking that led to the theory of relativity. Below is a basic summary of Wertheimer's analysis of productive thinking, which at its core has the goal of "structural improvements";

1. Structural analysis that identifies the "trouble regions," "gaps," "disturbances," and "superficialities" that stand in the way of achieving the structure;
2. Addressing the gaps in the structure;
3. Since the parts interact, changing a part of the structure will necessitate a restructuring of the whole.[20]

Major restructurings are far more advanced and might win you a Nobel Prize, but smaller restructurings are proposed all the time. As Max Wertheimer wrote in *Productive Thinking*, the goal is "looking for structural rather than piecemeal truth."[21]

So, to clarify, while proposing that the lecture still has a place in higher education, I do not mean to imply that "active learning" is invalid.[22] Active learning continues to be a part in even the most traditional classroom in the form of papers and homework assignments. A student "constructing their own learning" is certainly valuable and is the flip side of memory being a "reconstruction" rather than a "reproduction," in the words of Frederick Bartlett.[23] Students need to develop an independence in their thinking and research, construct their own schema, and look for improvements and weaknesses in current theories and ideas. They need to formulate an argument, present their findings, face challenges, and defend them. They can only become skilled at this by *doing*. As students become more advanced, professors can reasonable expect more from them in this regard. And certainly, there can be advantages and a convenience to students engaging in active learning during class time with the professor present. However, students are not only in need of developing critical thinking skills but are equally in need of examples and organization provided by the professor. To refuse students of this is akin to learning to play the cello without ever listening to a professional cellist. After all, what better role model for students then their own professors?

Lee Cronbach (1916–2001), a significant figure in the field of psychological testing and educational psychology, provided valuable recommendations for the introduction of new material in the classroom:

> The person hearing or reading a complex communication about unfamiliar material cannot store the incoming pieces of the message efficiently unless he perceives the general direction the argument is taking. Without that overall conception, he has to store sentences one at a time, with consequent distortion and rapid loss. If properly oriented to the communication at the onset, he calls upon his background of relevant images and uses them to interpret the new

> communication. It is well, then, to begin an instructional communication with some kind of initial orientation . . . . When the nature of any detail is determined by the whole—and this is true of machines, scientific theories, literature, and history—then understanding of the part must wait on comprehension of the whole. Before presenting details, it is wise to give a general overview so that each part will be more meaningful . . . . The organization that seems to be the most useful generally is the whole-then-part approach.[24]

While Cronbach does not specifically mention it, implied is a dynamic interplay between the parts and the whole in which the instructor points the attention of the students sometimes to the whole, at other times to the mastery of the parts, and still at other times to the meaningful relations between the parts. As a result, I prefer to think of the rhythm of teaching as a pendulating whole-part-whole design. This approach reflects the way our own LTM stores information and sets the student on the path to a deeper comprehension of the subject matter.

This approach can be applied to the music history lecture in any number of ways. In music history classes, I typically present a broad overview of the topic at the beginning of the semester. One possibility is to paint in broad brush strokes the musical periods to be covered, drawing comparisons and contrasts between them. In this case the "whole" is musical time periods and would be appropriate for a class focusing on a history of musical styles. Another option is to begin with broad musical characteristics that apply to all cultures, time periods, and styles, emphasizing the similarities of musical expression.

Here the "whole" is music itself and might be appropriate in a world music class or in a class that will emphasize the similarities between time periods. Still another example is to provide an overview of extra musical events that took place during the time period in question, making references to the influences these historical events had on music. This would be an even larger "whole" than the previous two and would emphasize how music reflects and interacts with socio-political events, intellectual movements, religious beliefs, and broader culture. Another possibility is an overview of the lineage of compositional influence, emphasizing the family tree in music history, showing as in a domino effect the organic emergence and connectedness of composers through musical time periods. In this fourth example, the whole is an unfolding chronology of influence, not unlike a slowly developing minimalist work by Phillip Glass, yet with sudden changes interjected. This would be well suited to a chronological approach to music history and its advantage would be to show the connectedness between seemingly disperate styles. An incredible array of options is available to the professor in deciding the big picture to which the subdivisions will relate throughout the semester.

In music appreciation classes, I also attempt to relate new information to other topics the students are likely familiar with or may connect to, allowing them to attach meaning to an unfamiliar musical tradition. For example, I have compared Aaron Copland to John Mellencamp as examples of "heartland" composers. I have also compared 12-tone music to *No wave* avant-garde rock music in their use of dissonance and atonality. These comparisons are not perfect but allow students to make meaningful associations and may provide shortcuts to LTM.

Music theory courses, which traditionally tend to be lighter on lecture and more focused on in-class learning, can benefit greatly from a pendulating whole-part-whole approach. Much music theory curriculum has tended towards a part-then-whole approach, starting with the basic elements of theory such as the staff, clefs, pitch, scales, circle of fifths, and key signatures followed by examples from the literature to illustrate the points. Another approach is to start with compositional scores from which the same components mentioned above are extracted but are now seen in a musical context. Pieces can be chosen to illustrate differing scales, clefs, key signatures and so on.

This is especially helpful for vocal or music theatre students who do not always have the same background in reading music as instrumental majors. It is recommended to listen to the piece first or perhaps sing it, if appropriate. After moving from the composition to focusing on mastering the parts, returning to the score can then illustrate how the parts interact, providing opportunities for analysis. This basic instructional rhythm has also been suggested by Swanson and Law.[25] A pendulating whole-part-whole approach can be utilized for any level of music theory. The focus on musical context also prevents music theory from becoming a sterile exercise in the minds of the students and demonstrates that all analysis, in the end, is a musical, interpretive exercise.

In this chapter I have sought to persuade the reader, without negating the role of independent learning, that expert instruction in the form of the lecture has a vital place in higher education. If done correctly, students can be presented with context, organization for new material, and a model for thinking, hopefully being inspired in the process. Emphasizing a pendulating whole-part-whole approach reflects the way our LTM stores information and provides opportunities for the professor to aid students in the development of schema. A whole array of possibilities is available that invite creativity on the part of the professor.

NOTES

1. P. A. Kirschner, J. S. Sweller, and R. E. Clark, "Why Minimal Guidance During Instruction Does Not Work: An Analysis of the Failure of Constructivist, Discovery, Problem-Based, Experiential, and Inquiry-Based Teaching." *Educational Psychologist* 41/2 (2006): 77.
2. M. Haselgrove, *Learning: A Very Short Introduction* (Oxford: Oxford University Press, 2016), 22.

3. E. Tulving and F. I. M. Craik, eds., *The Oxford Handbook of Memory* (Oxford: Oxford university Press, 2000), 114.

4. L. J. Cronbach, *Educational Psychology, 3rd edition* (New York: Harcourt Brace Jovanovich, Inc., 1977), 328.

5. C. H. Stilwell and A. B. Markman, "Schema-driven Memory and Structural Alignment." *Proceedings of the Annual Meeting of the Cognitive Science Society* 25/25 (2003): 1122.

6. J. K. Foster, *Memory: A Very Short Introduction* (Oxford: Oxford University Press, 2009), 65.

7. Foster, *Memory: A Very Short Introduction*, 8.

8. J. Johnson, *Out of Time: Music and the Making of Modernity* (Oxford: Oxford University Press, 2015).

9. H. Gertrude, "The Difficulty Reduction Tendency in Perception and Problem Solving." *Journal of Educational Psychology* 32 (1941): 305–13.

10. L. J. Cronbach, *Educational Psychology* (New York: Harcourt, Brace and Company, Inc., 1954), 279.

11. M. Wertheimer, *Productive Thinking* (New York: Harper & Brothers Publishers, 1945), 191.

12. G. M. Heider, "More about Hull and Koffka." *American Psychologist* 32 (1977), 383.

13. Tulving, and Craik, eds., *The Oxford Handbook of Memory*, 114.

14. W. G. Chase and H. A. Simon, "Perception in Chess." *Cognitive Psychology* 4/1 (1977): 55–81.

15. Tulving and Craik, eds., *The Oxford Handbook of Memory*, 115.

16. Foster, *Memory: A Very Short Introduction*, 64.

17. J. D. Bransford and M. K. Johnson, "Contextual Prerequisites for Understand: Some Investigations of Comprehension and Recall." *Journal of Verbal Learning and Verbal Behavior* 11 (1972): 717–26.

18. Ibid., 722.

19. Ibid., 724.

20. Wertheimer, *Productive Thinking*, 190–1.

21. Ibid., 191.

22. I only object to what is implied by the phrase "active learning"; namely, that listening to comprehend is properly described as "passive."

23. Foster, *Memory: A Very Short Introduction*, 9.

24. Cronbach, *Educational Psychology, 3rd edition*, 522–3.

25. R. A. Swanson and B. D. Law, "Whole-Part-Whole Learning Model." *Performance Improvement Quarterly* 6/1 (1993): 43–53.

WORKS CITED

Bransford, J. D. and M. K. Johnson. 1972. "Contextual Prerequisites for Understand: Some Investigations of Comprehension and Recall." *Journal of Verbal Learning and Verbal Behavior* 11: 717–26.

Chase, W. G. and H. A. Simon. 1977. "Perception in Chess." *Cognitive Psychology* 4/1: 55–81.

Cronbach, L. J. 1954. *Educational Psychology*. New York: Harcourt, Brace and Company, Inc.

Cronbach, L. J. 1977. *Educational Psychology, 3rd edition*. New York: Harcourt Brace Jovanovich, Inc.

Foster, J. K. 2009. *Memory: A Very Short Introduction*. Oxford: Oxford University Press.

Gertrude, H. 1941. "The Difficulty Reduction Tendency in Perception and Problem Solving." *Journal of Educational Psychology* 32: 305–13.

Haselgrove, M. 2016. *Learning: A Very Short Introduction*. Oxford: Oxford University Press.

Heider, G. M. 1977. "More about Hull and Koffka." *American Psychologist* 32/5: 383.

Johnson, J. 2015. *Out of Time: Music and the Making of Modernity*. Oxford: Oxford University Press.

Kirschner, P. A., J. S. Sweller, and R. E. Clark. 2006. "Why Minimal Guidance During Instruction Does Not Work: An Analysis of the Failure of Constructivist, Discovery, Problem-Based, Experiential, and Inquiry-Based Teaching." *Educational Psychologist* 41/2: 77.

Stilwell, C. H. and A. B. Markman. 2003. "Schema-driven Memory and Structural Alignment." *Proceedings of the Annual Meeting of the Cognitive Science Society* 25/25: 1122–7.

Swanson, R. A. and B. D. Law. 1993. "Whole-Part-Whole Learning Model." *Performance Improvement Quarterly* 6/1: 43–53.

Tulving, E. and F. I. M. Craik, eds. 2000. *The Oxford Handbook of Memory*. Oxford: Oxford University Press.

Wertheimer, M. 1945. *Productive Thinking*. New York: Harper & Brothers Publishers.

Chapter 13

The Religious Lecture

Synthesizing Persona and Narrative Immersion

Paul Krause

Since the late nineteenth century religious studies has been conflicted by multiple, often mutually agonistic, means of study. The rise of biblical criticism, and historical criticism, has led to the religious lecture having limited theological or even literary explanation (in the sense of focusing on literary patterns, archetypes, and narrative structure). The lecture is Hegelian in nature and uses religious works as an insight into the past which the lecture seeks to reconstruct.[1] The counter to historical criticism was the new fundamentalism; the (confessional) religious lecture as the teaching moment where the professor affirmed doctrinal orthodoxy and explained it to the students. More recently, however, the importance given to the oral embodiment of religious texts has led to new interests in a synthesis of the historical critical-biblically inerrant dialectic. Because the lecture is intrinsically oral, and the lecture is itself a storytelling medium, the religious lecture is uniquely suited to be the synthesis of insight into the past while remaining "true" to the ancient nature of religious stories and the truths they embody.

This chapter deals with the oral nature of the lecture, the personal nature of the lecture, and how the oral and personal nature of the lecture perfectly embodies an integral aspect of the study of religion. Drawing on evolutionary psychology, the social understanding of human nature, and synthesizing the historical critical and scripturally inerrant dialect that has dominated religious studies and the religious lecture for the past century, I will argue that the religious lecture is the purest embodiment of "narrative immersion" through persona and textual signification.

The problem that the religious lecture faces is the question: What is the role the religious lecture aims to achieve? The religious lecture is a delicate subject given the dialectic between the lecture promoting the historical-critical method of higher textual criticism and the response to it through the confessionalist lecture. Indeed, notable institutions of higher education were fragmented on the nature of the religious lecture. Princeton University, founded by American Presbyterians, experienced the equivalence of denominational schism as Professor John Gresham Machen, among others, left to found Westminster Theological Seminary in protest of the acceptance of higher criticism by the Princeton faculty. Similar schisms occurred at other institutions of higher education.

The study of religion situates itself on a fragile precipice. It is, on one hand, threatened by becoming absorbed to history—the religious lecture nothing more than a specialized type of historical lecture with a focus on the phenomenon of religion. On the other hand, the religious lecture is threatened by being held in contempt by other academic disciplines for being confessional—a non-scientific enterprise of little value beyond doctrinal enforcement. To be taken seriously the religious lecture abandons religion to historical. To save souls and uphold the truth, the religious lecture is subservient to creeds and confessions. Both outlooks are in conflict and miss significant opportunities for rich pedagogical immersion and participation.

Apart from the well-founded connection between orality and literacy,[2] literacy is linked with orality and the decline of humanity's oral core has negative consequences to literacy. The loss of orality precipitates the loss of literate culture which sprung from the oral well. The integrity of the spoken word, spoken language, is essential to a literate culture.[3] Orality and literacy are not mutually opposed but are in a sort of symbiotic relationship to each other. Moreover, it has also been said that the "Orality-literacy dynamics enter integrally into the modern evolution of consciousness toward both greater interiorization and greater openness."[4] If true then orality has a profound impact not merely on socialization but also moralization.

It is well known and accepted that humans are social creatures.[5] Given this reality and the reality of the oral-literate dynamic, the oral lecture is not only the best medium of communicative education, it also serves contingent social and moral purposes. What neuroscientists and neurobiologists call "narrative immersion" has a connection to oxytocin which influences moral concern for others.[6] Storytelling, that most intrinsically oral construction, leads to greater engaged social behavior. Moreover, narrative immersion is the key any successful lecture which depends on the dialectical dynamism of speaker and listener (teacher and student).

There seems to be a disconnect between the "boring lecture" and the narrative immersion of storytelling which enthralls the mind and has productive and positive ramifications for human behavior and imagination. Joseph Campbell studied the art of the successful myth—the hero's journey or hero's quest (the "monomyth")—and found standard archetypes that link the great stories together through time and culture.[7] Religious stories follow these narrative archetypes. As such, people are, in a way, wired to gravitate to such stories and become immersed in them. Identifying with the hero, wishing the hero to succeed, or to become one with the story—to become part of the unfolding story. Narrative immersion invites participation rather than remain aloof and isolated (table 13.1).

The key to any success in a lecture is to grip the audience into participation rather than build a barrier between speaker and listener. In this manner the lecture seeks, in a very Kantian way, to achieve the apperceptive unity between the I and Not-I; between the speaker and listener, between the teacher and student. And bridging this connection is the key to the lecture's success. Inability to make this connection, to achieve narrative immersion into the content of the lecture, is what leads to the disconnect between lecturer and student. The active-passive lecture dichotomy is tied to narrative immersion

Table 13.1 The Hero's Journey

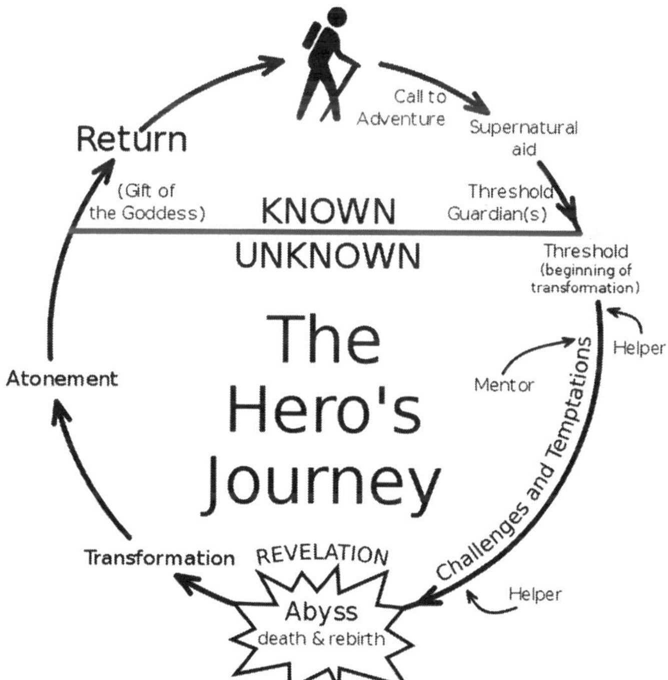

Above: A depiction of Joseph Campbell's "monomyth" of the Hero's Journey, charting the outline of the adventure or journey that all great stories use and which the psyche is predisposed to gravitating to. Given the centrality of the Hero's Journey to religious stories, the religious lecture is best suited to immerse students into the narrative of Hero's Journey.

and whether the recipients—students or listeners—immerse themselves in the narrative which is the truest form of participation: To become one with the lecture, story, or narrative.

This embodiment of orality is part of what Walter Ong called second orality.[8] Second orality is the, "more deliberate and self-conscious orality, based permanently on the use of writing and print."[9] Techne, then, is integrally related to the success and continuation of orality in its secondary forms: literature, note-taking, and now even the virtual lecture. Narrative immersion—that active participation on part of the student—is bound up with second orality. For reception of orality and transmission of orality into its secondary forms: literature, note-taking, learning, is the second orality that Ong speaks of. In this manner successful narrative immersion translates into the successful development of second orality, which is manifested in note-taking and memory as it relates to students.

Enthusiasm and passion, on the part of the teacher and the student, may have an impact on the successful transmission of knowledge in the lecture. However, the danger with extensive enthusiasm from the teacher is to be immersed with the persona of the teacher rather than the content of the lecture. Persona immersion may build close relationships with people, lead to a student liking the teacher, but serves as a barrier to the participatory lecture through narrative immersion. Persona immersion is also an unintentional barrier to oral signification and, therefore, narrative immersion is what the lecture primarily aims to achieve with persona immersion being secondary.

Considering the philosophy of language offered by Augustine of Hippo, to be wrapped up with persona rather than speech is to confuse the signifier for the thing—for all learning, Augustine says, is done by way of signs.[10] Part of Augustine's philosophy of language, abstracted from his broader metaphysics, is the distinction between signifier and signified thing wherein he argues that speech is the chief sign in the world.[11] The teacher is the conduit of signification and not the thing to become enamored with; the teacher is the signifier needed for learning. To confuse the signifier for the signified poses dangers; as it relates to the lecture this is the fall for persona immersion where the persona of the teacher becomes the embodiment of signification rather than the lecture material which stands at a distance because it is no longer that which is being signified. The spoken word is a medium to signify but is not the thing signified in-of-itself. Popular sayings like "what is the moral of the story" still bear out this reality: That stories signify, so too is that the case with the lecture.

The teacher needs to be aware of this (as do students). The lecture is a waltz. A grand song of life. It is an orchestra with many constitutive parts to it moving for a goal that can easily miss the mark by simple deviations. The task of the teacher is to be energetic and passionate—as any great storyteller must be—but not at the point of overwhelming the lecture. To draw students into mere persona immersion is just as counterproductive to the lecture's signifying nature as the uninviting lecture due to boredom. But the lecture cannot exist without the persona of the instructor. Much like how it cannot exist with narrative or oral form. Nor can the lecture exist without participants.

The dramatis personae of ancient Greece captured this reality in the plays which were performed which follow the same structure and embodiment that lectures do. The mask, persona, worn by the actors was a mark of signification. The persona was meant to signify something beyond the mere actor. While the lecture may not contain in it the multitude of dramatis personae that ancient plays did, the lecture still contains the principal speech actor (the teacher) and the audience (the students) in which the teacher is the conduit of signification for the students. A successful play, like a successful lecture, is that delicate waltz between both ends of this dialectic.

This is not to criticize persona immersion per se. People have an instinct to gravitate to heroes. Persona is contingent to the narrative as hitherto stated. But the balance must be struck between the instructor as signifier instead of the instructor as signified.

The religious lecture, and the very nature of the study of religion, occupies a unique place in this respect. Persona is very much a central aspect to religion. Religion has gods, heroes, prophets, peoples, in other words, personas. But what persona signifies is more important. Narrative immersion is tied to persona. No one confuses what heroes and gods and teachers represent in religious stories; for the heroes, gods, and teachers are the embodiment of signification and teaching or instruction. This is one of the reasons why the "boring teacher" is unable to capture the attention of students and bring them into narrative immersion. There is a symbiotic relationship between teacher and student, between persona and narrative immersion, which comes together in the lecture to produce the most effective learning form for humans. No one is immersed by a boring story.

The human psyche is already predisposed to certain narrative archetypes and functions from the long history of oral evolution.[12] While motivation is always helpful, motivation on part of the students for learning is not necessarily the key to a successful lecture. As mentioned, the ability to bring the listener into the depths of narrative immersion is the key to a successful lecture or speaking engagement irrespective of the topic. Psychologically, people are used to, and immersed in, the hero archetype.[13] The hero archetype, formalized in the "hero's quest," is foundational to literature: secular and religious.[14] The human psyche loves a good story. The psyche desires immersion into the narrative, and the psyche is drawn to heroes, adventures, and sacrifices—the "total human story" which naturally captures the psyche and draws its attention to the meaning of what is being communicated through the medium of orality.[15]

The human mind also seeks patterns in stories, "As humans evolved from their anthropoid ancestors, and the size of their cerebral cortex expanded, novel pattern processing capabilities emerged."[16] This instinct for patterns has, therefore, developed from humanity's oral past, "Storytelling allowed organization, structure, and the meaning-making of internal memories shared by a collective experience of living. This collective wisdom enhanced and enabled survival and formed the foundation of community."[17] As it relates to narrative immersion, the psyche seeks familiar narrative patterns and structures like the ones hitherto mentioned to latch onto and become immersed in. It is precisely because religious texts share the fundamental basics of oral archetypal embodiment tied to the prospects of narrative immersion which allows for the possibility of the religious lecture to embody narrative immersion so successfully. Religious texts, and religious stories, captured that transition from first orality to second orality while retaining and cementing that "shared . . . collective experience of living."

If the teacher can reawaken the psyche's search for patterns, signifying how texts and stories related to religion are the genus of literary organization, structure, and meaning-making, the teacher will have a successful participatory lecture. This returns us to the question as to what is to be signified, or promoted, in the religious lecture: The historical-critical or the confessional? In this respect that incessant concern over this division is incidental to the more fundamental core of the religious lecture: How to immerse the student in the narrative of the religious lecture? The problem with the historical-critical lecture method is it fails to draw on the rich archetypal keys waiting to be (re)awoken in the psyche which can draw the student into narrative immersion. Likewise, the confessionalist lecture suffers the same problem; thus, turning the lecture and its receptivity into a game of intellectual ascent.

The religious lecture, to be successful, should move beyond the singular overburdensome Hegelian presupposition and return to textual analysis of literary archetypes and narrative structures. Religious texts were among the first written and codified literary works; if not explicitly religious or regarded as more poetic mythology, even poetic mythology embodies religious themes along with the literary archetypes and narrative structures conducive to narrative immersion.[18] The religious lecture has, already built into the texts being studied, all the archetypes and literary structures the psyche is predisposed to liking and being immersed into.

The problem with the historical-critical method of the religious lecture is that it strips away the very wealth of images and possibility for narrative immersion which all successful lectures exude. The compensatory fantasy that Carl Jung spoke of, where the world was presented in the images of the unknown in which images serve as signifiers of that which was unknown,[19] is what the historical-critical method does away with (in its bid to "demythologize"). And the historical-critical method seems unable to recognize the human truths contained in myth in favor of post-Hegelian disenchantment.

The historical-critical religious lecture stacks the odds against itself by removing itself from all the gifts of religion and religious texts suitable for narrative immersion and substituting these with ungrounded hypotheses like the Q-source, dry explanations of language comparison between manuscripts to prove scribal editing, and intertextual analysis and cross textual comparison for dating. Even where the historical-critical method attempts to explain ethics and metaphysical or epistemological themes, it resorts to a sort of tabula rasa presupposition which is equally embodied in the style of the historical-critical lecture. It assumes a blank slate on the part of students (or need to establish a blank slate by "unlearning your church Bible study") and the religious text in question; the "pure" text to be (re)constructed through historical-critical analysis which is then thrust onto the students in a top-down imposition in the lecture which involves little participation and relies on forced engineering. This top down imposition through forced engineering equally reveals the blank slate mentality behind the historical-critical method. There also remains in the historical-critical approach an implicit inheritance of the Protestant doctrine *sola scriptura* too, even if historical-critical methodology rejects textual inerrancy and infallibility.

The confessionalist lecture suffers from the same problems of the historical-critical approach but from the opposite pole. It too, in a way, assumes a blank slate to the catechumenal mind which must be filled with the correct intellectual beliefs concerning theology and textual interpretation. The confessionalist lecture signifies the standards of historical creeds and confessions at the expense of literary archetypes, story structures, and the narrative of the hero's journey. Like the historical-critical lecture, the confessionalist lecture handicaps itself by not utilizing the wealth of second orality waiting to be tied back to original consciousness which has been synthesized to the second orality of the written word and paper as the construction of first orality. It is that synthetic relationship between first orality (original oral consciousness) and second orality (the deliberate and self-conscious effort at codifying first orality) that provides the basis for narrative immersion.

Given that the traditional means of delivering the religious lecture do not take advantage of the rich literary and narrative archetypes available to it, a new literary exegetical approach—one that puts the historical-critical or confessionalist lecture disposition secondary—focusing on literary archetypes, story structures, and the hero's journey, and how these literary constructions signify basic human themes of struggle, sacrifice, and heroism, would likely lead to successful immersion into the religious lecture. Moreover, it is the lecture that successfully draws upon these predisposed narrative structures and themes which would successfully immerse students into the lecture.

The religious lecture has the unique opportunity to embody the fullness of narrative immersion in retelling the song of life of early ancestors. In this way the religious lecture can also teach us something about our past as the historical-critical method aims at and why sacrifice, life, and death factor so prominently in religious confessions as the confessionalist method aims at. The lecture is also the modern embodiment of this primordial storytelling past, surrounded around the hearth—now the classroom—in which a body of people gather for a story. For the lecture is a story with a beginning, middle, and end which intends to signify something to the audience. As mentioned, the historical-critical lecture attempts to signify how texts were constructed and why. The confessionalist lecture tries to signify the truth and relevancy of texts and why they are essential to certain faith traditions. Signification runs through the heart of all lectures; but not all lecture methods draw effectively on archetypes and literary structural patterns conducive to narrative immersion.

If the religious lecture still aims at showing how texts and the stories contained therein came about, a more effective and still historically accurate—perhaps in a meta sense—of how these texts and stories came into being would be through a recovery of the literary focus of religious works and the transmission from first orality to second orality. The religious lecture has the unique wealth of character archetypes, storylines, and structures that the psyche is predisposed of gravitating toward and immersing itself in. This activates the participatory nature of the lecture between speaker and listener; teacher and student.

The religious lecture that highlights the literary structure, the character archetypes, and the signification of the literary structure, journey, and hero, would not only be a lecture and educational environment engaged in good textual criticism but also double as the ideal conduit for narrative immersion in the form of the lecture. That religious stories and heroes functionally embody the oldest of cultural and literary archetypes, structures, and tropes, the study of religion and the teaching of religion through the extraction of second orality back to first orality in the lecture form may allow for a great reintegration of persona and narrative immersion. All lectures rise and fall on the ability to signify for students the narrative form and patterns that the psyche can be immersed in.

The personal lecture, with the teacher as the persona signifier the more important meaning of the stories, is still the most suitable manner for teaching. The personal and relational nature of humanity ensures that the most effective means of personal instruction and learning is still the time-tested and evolutionarily embodied reality of the oral lecture. Narrative immersion remains the strongest means of education and participation. And the religious lecture, with its wealth of literary archetypes to draw on and emphasize in the lecture, is most aptly suited for this task.

NOTES

1. John T. Noonan, "Hegel and Strauss: The Dialectic and the Gospel," *The Catholic Biblical Quarterly* 12/2 (1950): 136.
2. See Walter J. Ong, *Orality and Literacy* (New York: Routledge, 2002).
3. Anthony K. Webster, "Keeping the Word: On Orality and Literacy," *Oral Tradition* 21/2 (2006): 295–324
4. Ong, *Orality and Literacy,* 175.
5. See Max Tomasello, "The Ultra-Social Animal," *European Journal of Social Psychology* 44 (2014): 187–94.
6. Paul Zak, "Why Inspiring Stories Make Us React: The Neuroscience of Narrative," *Cerebrum* 2 (2015).
7. See Joseph Campbell, *The Hero with a Thousand Faces* (Novato: New World Library, 2012).
8. Ong, *Orality and Literacy,* 133.
9. Ibid.
10. Augustine, *De Doctrina Christiana* (Hyde Park: New City Press, 2015), 1.2.2.
11. Ibid., 2.3.4.
12. Ellen Spolsky, "Archetypes Embodied, Then and Now," *Poetics Today* 38/2 (2017): 317–39.
13. Campbell, *The Hero with a Thousand Faces,* 1–18.
14. Ibid.
15. Christopher Vogler, *The Writer's Journey* (Saline: McNaughton & Gunn Inc., 2007), 26.
16. Mark P. Mattson, "Superior Pattern Processing is the Essence of the Evolved Human Brain," *Frontiers in Neuroscience* 8/265 (2014).
17. Monika Reis, "Once Both In and Out of Time: Language, Storytelling, and Transformation," *Assisi Institute Journal* 1/1 (2014): 114.
18. Markus Altena Davidsen, "Fiction and Religion: How Narratives About the Supernatural Inspire Religious Belief – Introducing the Thematic Issue," *Religion* 46/4 (2016): 489–99.
19. Carl Jung, *The Structure and Dynamics of the Psyche* (New York: Routledge Press, 1960), 60.

WORKS CITED

Augustine. 2015. *De Doctrina Christiana*. Hyde Park: New City Press.

Campbell, Joseph. 2012. *The Hero with a Thousand Faces*. Novato: New World Library.

Davidsen, Markus. 2016. "Fiction and Religion: How Narratives about the Supernatural Inspire Religious Belief – Introducing the Thematic Issue." *Religion* 46/4: 489–99.

Jung, Carl. 1960. *The Structure and Dynamics of the Psyche*. New York: Routledge Press.

Mattson, Mark. 2014. "Superior Pattern Processing is the Essence of the Evolved Human Brain." *Frontiers in Neuroscience* 8/265. Available at https://www.frontiersin.org/articles/10.3389/fnins.2014.00265/full.

Noonan, John. 1950. "Hegel and Strauss: The Dialectic and the Gospel." *The Catholic Biblical Quarterly* 12/2: 136–52.

Ong, Walter. 2002. *Orality and Literacy*. New York: Routledge.

Reis, Monika. 2014. "Once Both In and Out of Time: Language, Storytelling, and Transformation." *Assisi Institute Journal* 1/1: 113–23.

Spolsky, Ellen. 2017. "Archetypes Embodied, Then and Now." *Poetics Today* 38/2: 317–39.

Tomasello, Max. 2014. "The Ultra-Social Animal." *European Journal of Social Psychology* 44: 187–94.

Vogler, Christopher. 2007. *The Writer's Journey*. Saline: McNaughton & Gunn Inc.

Webster, Anthony. 2006. "Keeping the Word: On Orality and Literacy." *Oral Tradition* 21/2: 295–324.

Zak, Paul. 2015. "Why Inspiring Stories Make Us React: The Neuroscience of Narrative." *Cerebrum* 2. Available at https://www.ncbi.nlm.nih.gov/pmc/articles/PMC4445577/.

Chapter 14

Creating a Socratic Lecture Space in Religious Studies

Paul Corey

This chapter will make two main points regarding lectures in religious studies (RS) courses at non-confessional universities and colleges. First, it will reimagine the lecture space and defend a type of oral engagement modelled on the Socratic approach.[1] This is a student-centred pedagogy, a form of active learning, where students are engaged participants in truth-seeking dialogue, rather than being passive receptors of professorial monologue. Second, the Socratic Lecture Space, as I will call it, should have a prominent place within RS, particularly in undergraduate courses studying comparative religion. This will help to bridge what has been called "the great divide" between lecturers and students in RS courses. However, the Socratic emphasis on critical dialogue increases the possibility of disagreement and confrontation, which needs to be both welcomed and directed by the professor.

THE GREAT DIVIDE

In 2008, Barbara Walvoord published findings that identified what she called the "great divide" in RS classrooms.[2] Walvoord's study involved 533 instructors and 12,463 students in introductory religion and theology courses at 109 universities and colleges throughout the United States.[3] Both instructors and students were asked about their goals for learning. The most frequently chosen goal cited by professors was "critical thinking," which Walvoord claims takes three main forms: (1) "analyzing the historical, cultural, linguistic, literary, political and social contexts of religious beliefs and practices"; (2) "critically evaluating arguments and points of view"; and (3) "constructing one's own arguments about theological and religious issues, relying on reason, evidence, and logic."[4] The first type of critical thinking, concerned broadly with historical context, has been of particular interest to RS scholars over the past few decades. Often falling under the broad category of "Critical Theory," it explores how the contingencies of historical context shape both religion and the academic discipline of RS. In other words, it examines both religion and RS as subjective constructs—constructs that privilege certain discourses, practices, and peoples over others—not as universal accounts of truth.[5] Critical thinking, in this sense, is drawing students' attention to the linguistic, historical, and cultural contexts of religious beliefs and practices, and deconstructing the unreflective assumptions within the academic discourse of RS itself. Essentially, the goal is to get students to consider *in whose interest* a discourse serves, and the power dynamics established when a discourse becomes authoritative. This means that there is no such thing as a neutral approach to the study of religion. Even the term "religion" itself is problematic, since it emerged within a Western cultural context, with its own assumptions about what constitutes "religious" phenomena.[6] The goal of many RS professors is to draw attention to such implicit biases and encourage self-reflexivity in the study of so-called religion.

In Walvoord's study, student goals in RS courses were somewhat different. First, students wanted to learn basic factual information about various religions, including their own; but also, and more noteworthy, they hoped to "develop their own spiritual and religious lives."[7] According to Walvoord, students understand "spiritual development" to be connected to "big questions" such as: "Who am I? Why am I here? What is the good life and how do I achieve it? What is truth?"[8] Students in the study were expecting that their RS courses would engage such questions. Some students said they welcomed a challenge to their own beliefs, whereas others hoped that the course would strengthen their existing faith. Whatever the reasons, students were seeking some type of spiritual guidance from their RS courses.[9]

The Critical Theory approach to RS would seem to be at odds with the "big questions" approach desired by students. Since religious and academic discourses are reducible to constructions that create subjective knowledge/power hierarchies, then any answers to such questions are hopelessly biased; indeed just asking the questions themselves comes with a host of unexamined assumptions that privilege certain categories, norms, and individuals over others. Thus, the critical thinking espoused by most professors at postsecondary institutions is ostensibly in tension with, or antithetical to, the spiritual concerns of students. Nowhere is this more apparent than in RS lectures, where professors deliver academic soliloquies behind the podium, focused on their own self-reflexive critical analyses, while alienating students, who may be left wondering how any of this addresses their deepest questions. In general, faculty in Walvoord's study expressed reluctance to address students' spiritual needs, either out of conviction that professors must stick strictly to academic research, or out of fear that spiritual discussions could lead to uncomfortable moments in the classroom. Some professors suggested that "spiritual development" was an unstated "sub rosa" goal, or a happy-but-unintended consequence of the course, but nothing that could be stated as an official learning outcome.[10]

In response to the study's findings, Walvoord investigates ways to bridge the great divide. For some, this effort is misguided. Christopher Cotter and David Robertson, proponents of the Critical Theory approach, write that Walvoord "problematically assumes that students' spiritual development is within the purview of RS."[11] This implies that RS professors should simply stick to academic critical analysis and exclude students' spiritual concerns, while, ironically, lecturing on subject matter that deals expressly with these very concerns. This irony forces RS professors to confront questions about the goals of higher education, particularly in the liberal arts. The liberal arts were traditionally conceived as studies that transform students into *free* (i.e., "liberal") citizens by cultivating the virtues necessary for free choice—the abilities to think critically, speak rationally, and engage civilly. In other words, the liberal arts immersed students in the examined life, through which they could flourish as thoughtful citizens. Critically examining religious beliefs and practices would seem to be a crucial aspect of the examined life. The RS class, therefore, has the potential to affect students at the deepest level, since it may touch upon their deepest religious—or anti-religious—commitments. To address students' spiritual concerns is not synonymous with "strengthening their faith," but rather encouraging students to assess their opinions and practices, and consider other ways of understanding and living in the world. In this sense, critical thinking is not only a consideration of historical context but, in accord with Walvoord's three-part definition, also *evaluates opinions*, and *constructs new arguments* based on reason, evidence and logic.

In her study, Walvoord compiles the pedagogical strategies of those she calls "Highly Effective Faculty" who cultivate critical thinking while "making space for students' religious and spiritual development." Such faculty create spaces that are safe from: (1) "teacher and classmate ridicule or denigration"; (2) the "pressure to commit to a particular religious tradition"; and (3) the "fear that it is wrong or dangerous to question religious beliefs." At the same time, such faculty create "voices" in a classroom where students are expected to articulate their own beliefs at certain moments, and to suspend them at others. One of these voices is the "committed critical thinker: stating one's own beliefs as arguments, and defending them."[12] The figure of Socrates stands as the ultimate example of such a critical thinker. But to truly adopt a Socratic approach would mean transforming the lecture from a space that is primarily monologic to one that is primarily dialogic.

THE SOCRATIC LECTURE SPACE

Norm Friesen provides an analysis of how the university "lecture" has survived in some form or other for nearly a millennium, notwithstanding changes in culture and technology that seem to threaten its relevance. Its stubborn persistence is due to its irreducible quality as a live meeting point or "interface," where the lecturer is essentially a medium or guide who connects novice students to advanced realms of knowledge—something that cannot be easily replaced by impersonal technology. Friesen identifies three types of lecture spaces over the course of Western history: (1) the medieval lecture as a "reading," where the lecturer recites an authoritative text to be taken down word for word by students; (2) the Romantic modern lecture as "sage from the stage," where the lecturer speaks without a prepared authoritative text but is himself the authority—the "Authorial Spirit"—who inspires students to create knowledge; and (3) the postmodern lecture as technologically enhanced performance, where the lecturer both uses, and accommodates herself to, electronic and digital media for effect.[13] Whatever the type of lecture, the lecturer in all three cases is delivering an authoritative monologue. There have been criticisms in recent years about the monologic qualities of the lecture, which put the lecturer front and center, and increasing efforts to make lectures more student-centred. As part of efforts to reimagine the lecture space, there has been a renewed interest in the "Socratic Method" as a classroom pedagogy.[14]

The Socratic Lecture Space (SLS) envisions the lecturer as, what I will call, a Daemonic Interlocutor, rather than as a reader, authorial spirit, or performer. The word *daemon* is important in Plato's dialogues, where Socratic pedagogy was first presented. In Plato's *Symposium, daemon* denotes a "spirit" between the imperfections of mortal life and the perfections of immortality.[15] Socrates practices philosophy—the "love of wisdom"—which is daemonic, caught between ignorance and wisdom, but desiring the latter; thus, the philosopher, like the lecture space, is an interface. But the philosopher, unlike the Romantic "authorial spirit" does not possess wisdom, and hence is not a sage full of wisdom who gives it to others; she is an erotic soul in pursuit of wisdom, and who tries to entice others to accompany her on this journey. Socrates' daemonic activity, for the most part, is not practiced through monologue; rather, it usually occurs through dialogue. In Plato's dialogues, Socrates practices *elenchus*: he asks his interlocutors questions, eliciting them to state their opinions on a range of topics, and then proceeds to cross examine these opinions through a series of further questions, to see if they are true, partially true, or false. Usually, an opinion will be shown to contain shortcomings or contradictions, leading to perplexity, and Socrates will ask the interlocutor to refine his understanding. This is what most people understand by the "Socratic Method." But Socrates does not just practice *elenchus*. Occasionally he uses myths and images to approach truth in other ways, suggesting *elenchus* alone is not enough—that it needs to be supplemented by narrative and the imagination. In Plato's dialogues, Socrates' mythmaking is the closest he gets to monologue, though it is always integrated within a larger discussion with others.[16]

If the lecture is to be reimagined as an SLS, then it is a space characterized primarily by dialogue. As with Socrates in Plato's dialogues, the lecturer remains central as a Daemonic Interlocutor. Though she must organize the topics of discussion for each class, the direction a class could take is open-ended. This means she will be usually working without a script, thinking on her feet as the dialogue takes unexpected twists and turns. The professor no longer exists at the front on a stage, hiding behind her monologue, fortified by a lectern, and taking the occasional question. The onus is on students to carry the discussion, prompted by the professor. As such, this breaks down the traditional division between "seminars" and "lectures," as the "lecture" would now also be a site of greater student involvement.

THE SOCRATIC LECTURE SPACE AND RELIGIOUS STUDIES

The SLS, particularly in an RS class, is less an environment of reverence than a place of critical questioning, which potentially leads to disagreement and conflict. This ostensibly goes against a prevailing orthodoxy in RS pedagogy: professors are to encourage not merely tolerance of all traditions, but respect and empathy as well, thereby creating a space of interreligious dialogue and mutual understanding. On the surface at least, it would appear that the SLS is not conducive to such a goal, with its emphasis on argument. Socratic pedagogy would seem to encourage potential confrontation—for example, if a professor or student were to point out how specific religious positions are untenable, or how different beliefs are irreconcilable, or how certain things done in the name of religion are unethical. Such arguments might appear disrespectful and offense may be taken.

Stephen Prothero, Professor of Religion at Boston University, states: "My students are good with 'respectful,' but they are allergic to 'argument.' They see arguing as ill-mannered, and even among friends they avoid it at almost any cost."[17] Often, attempts are made by either students or professors to diminish religious disagreements by saying "all religions are one," or different paths to the same goal, to which Prothero responds: "This sentiment, however well-intentioned, is neither accurate nor ethically responsible. . . . Faith in the unity of religions is just that—faith (perhaps even a kind of fundamentalism)."[18]

According to Alan Levinovitz, most comparative religious studies classrooms are dominated by either relativism or silence: relativism, because everything said or done in the name of religion is "respected"; silence, because students and professors are afraid to articulate their disagreements out of fear of being labeled racist, xenophobic, or culturally insensitive.[19]

Levinovitz, Associate Professor of Religion at James Madison University, writes "the essential purpose and unique objective of higher education requires a space where religiously intolerant confrontations can take place," confrontations that cause "existential challenges" to students' "innermost being and deepest commitments."[20] Against demands that lecture halls be intellectual "safe-spaces," where students should not encounter ideas they find offensive or personally challenging, Levinovitz argues that lecture halls should be unsettled spaces where confrontations are not only allowed but encouraged.[21] In other words, RS lectures should be reimagined as SLSs. As Levinovitz puts it: "Modern liberal education is supposed to combine the Socratic ideal of the examined life with a Millian marketplace of ideas. The product, ideally, is individuals who have cultivated intellectual virtues and ideas that have emerged victorious from communal debate."[22] To say that lectures are protected spaces where "intolerant confrontations" take place is not to say that lectures should descend into threats and violence. On the contrary, students learn the virtue of civil debate: how to withhold respect and intellectual assent for another person's beliefs and practices, while at the same time respecting the dignity of the person who defends these beliefs and practices. Students will *not* "agree to disagree"; they will *disagree*, but they will agree to basic civility. In the midst of disagreement, students must provide *reasons* for their assertions, not simply shout at each other with their unexamined biases. As such, genuine debate demands, in Levinovitz's words, "patience, open-mindedness, empathy, the courage to question oneself and the attempt to see things as others do."[23]

In the SLS, students are encouraged to debate the merits of various beliefs and practices. Are these beliefs and practices "true" in some sense (not necessarily literally) or are they outright false? Do they promote human flourishing or are they harmful? The discussion, of course, needs more nuance than simply dividing everything into "true or false," "good or bad." For example, a Socratic discussion could divide things up as follows: which beliefs and practices done in the name of religion can be *endorsed* by most people (the Golden Rule? charity for the poor?); which can be socially *accepted* but not practiced by most (dietary requirements? specific pilgrimages?); which can be socially *tolerated* but not accepted by most (biblical literalism? face veils?); and which are socially *intolerable* to most (mass suicide? terrorism?). Such divisions are by no means simple; what might be acceptable to some might be barely tolerable to others. But this is precisely the reason for such discussions: to see where a disagreement emerges and presses the issue.

If a student claims a belief or practice is demonstrably false or harmful, then she must defend her epistemological or moral intolerance with reasons. For example, a student cannot simply assert that wearing a niqab is wrong but must give reasons why it is detrimental to female flourishing; conversely, another student can argue that people opposed to the niqab are Islamophobic, but such a student should be encouraged to go further, articulating the reasons why wearing the niqab contributes to female flourishing. Of course, the lecturer must be sensitive to religious minorities in the class; for example, if only one student wears a niqab in class, this student should not feel centred out, bullied, or coerced by her non-Muslim classmates, nor should she be forced to defend herself. At the same time, if that student participates in class discussions of her own volition, and if she is willing to discuss the reasons behind the niqab, then this, and any emerging debate, should be welcomed. In the ethos of the SLS, defusing a debate by saying "different strokes for different folks" is an act of intellectual cowardice.

CRITICISM AND RESPONSES

The Socratic approach, similar to Critical Theory, wants people to examine their own presuppositions. This would also mean examining the assumptions of RS itself as a Western academic discourse. But there is a problem approaching all discourses (religious, academic) as simply subjective "constructs" that reinforce oppressive power hierarchies. All too often, such an approach descends into relativism, and enhances the great divide between students and lecturers. If the purpose of such deconstruction is to draw attention to the marginalized or oppressed who have suffered at the hands of these knowledge/power discourses, then that suggests that these discourses are measured by an ethical standard. Unquestionably all knowledge and practices are "constructs" shaped by language, culture, and historical context; but—and here is where the Socratic approach departs from Critical Theory—some constructs may be "truer" and more conducive to flourishing than others. For example, a self-reflexive discourse or practice that is conscious of its own hidden presumptions has greater epistemological and moral integrity than one that is not. Thus, the Socratic lecturer knows he or she is not wise, that is, he or she cannot provide a complete account of reality; at most, he or she can construct hypotheses or "likely constructs," some of which may be better than others.

That said, some might object to Socratic pedagogy insofar as it compels people to defend their beliefs or practices using words, and in this way is logocentric—focusing on texts, doctrines, and beliefs rather than practices and "lived religion." Over the past few decades, RS scholars have criticized this approach by deconstructing traditional categories used in the study of religion. Chief among these is the discourse associated with the "World Religions Paradigm"—a paradigm presupposed in most world religions courses. Cotter and Robertson argue that a "world religion" has traditionally been characterized as having a core "textual tradition," along with "a stress on orthodoxy rather than orthopraxy, and some degree of universal ambition." This discourse became hegemonic in RS, proclaiming itself as scientific (objective, universal, neutral), whereas in fact the entire discourse was shaped by Christian particularism in the nineteenth century. As such, it "prioritized 'belief' and 'doctrine' as preserved in texts as the *sine qua non* of 'religion'" rather than practices.[24] This means that scholars have traditionally either minimized or distorted those traditions that do not fit within this definition of religion—which is to say, most traditions outside of Christianity.[25]

Though the Socratic approach predates Christianity, the accusation could be made that it too privileges words, doctrines, and texts over practices and living religion. In response, it must be pointed out that Socrates himself is depicted by Plato as critical of written texts.[26] Socrates did not write books and was suspicious of their pedagogical value; his chosen mode of teaching was dialogue. In other words, the Socratic Method is not so much a textually based "methodology" in the modern sense, as a type of practice rooted in face-to-face oral engagement. And it is not only concerned with texts and doctrines, but also—and perhaps especially—with how people live their lives.

Nevertheless, the claim might still be made that the Socratic approach, with its emphasis on criticism, logic, and argument, has a "universal ambition" like a so-called world religion, and is thus "culturally imperialistic." This criticism is considered at length by Rebecca LeMoine. As she argues, the Socratic approach values "critical thinking and the input of students," thereby devaluing the "norm of respect for tradition and authority figures found in other cultures' educational approaches, such as the Confucian tradition."[27] Thus, Socratic pedagogy may commit imperialistic violence against traditions that do not value its approach. LeMoine, however, defends Socrates against this charge. First, as she points out, any teaching method will privilege certain normative values over others. For example, Confucian pedagogy favours authority, respect, and tradition; a teacher who practices ritual dance in class favours nonverbal discourse. No teaching method is neutral; all are constructs.[28] The Socratic approach favours questioning, discourse, and individual thinking; but it also is self-reflexive, aware of its own limitations, and can be subjected to critical assessment as well. This self-reflexivity may give it a preferred edge over other pedagogical norms, as it is *less* likely to be blindly imperialistic. Second, Socrates, when stating his own opinion, articulated provisional "hypotheses" that he favored while submitting them to critical assessment, but he did not impose "doctrinal truths" on his interlocutors. On the contrary, Socrates works within an interlocutor's "framework of thinking" to see where the thinking leads.[29] In an RS class, this means beginning with the premises accepted by a representative voice from a religious tradition and working from those premises. It does not impose foreign categories on the framework, but attempts to work within the framework itself. Furthermore, the Socratic approach cannot lead to proclamations such as "This religion (i.e. Buddhism) is better than this other religion (i.e. Christianity)." In accord with Plato's dialogues, Socrates considers *specific* opinions; no single voice or argument from within a religion—no matter how authoritative—can claim to represent everyone who identifies with that religion.

The problems identified with the "World Religions Paradigm" can be mitigated through what has been called "interactive interpretation."[30] Rather than organizing sections of a comparative religion course by "religion," it can be divided into themes, such as "pilgrimage," "sacrifice," "gender," and "war." The course designer must carefully choose materials from different traditions within each theme. Which themes and materials are privileged is a painstaking task, and professors need to be clear with students about what has been excluded. Nevertheless, the materials should be chosen from a wide range of sources and voices—orthodox and heterodox, powerful and disempowered, doctrinal and practical, textual and (as much as possible) oral. Some materials can focus on texts, beliefs, and doctrines, others might concentrate on practices, as well as the reasons practitioners give for those practices.

The effort to find the "reasons behind" certain practices and critically assess them may be objectionable for some, who might claim it violates the practices themselves. For example, it may not be possible to "critically assess" the reasons for the Hajj in Islam or the Kumbh Mela in Indian traditions; there is something nebulous about a phenomenon like "pilgrimage" that cannot be fully illuminated by the strong light of Socratic *elenchus*. But it is because of such intangibles that Socrates often resorts to narrative and myth in Plato's dialogues, to illuminate something that argument alone cannot quite capture. The widespread human need for pilgrimage might be better explained through storytelling and symbols, which can be done by the lecturer. However, to simply say that certain practices are beyond critical consideration, or that such considerations are inappropriate, is to renounce the examined life.

CONCLUSION

The point of this chapter is not to say that all RS lectures should be conducted Socratically. RS, as a field, encompasses numerous disciplinary approaches—sociological, anthropological, psychological, historical, biblical—and the Socratic approach is not always suitable given the academic demands of a specific discipline. Rather, the point is that the SLS should exist alongside other approaches in RS classrooms. There are, of course, other issues around how to accommodate the SLS, especially within large first-year classes that may not be easily conducive to dialogue. When RS departments design their curricula, a recommendation might be that the critical questioning of religion receives greater emphasis *after* first year, and that large introductory courses are more "Confucian" in approach, insofar as students are taught to understand and respect the religions they are studying. Given that most students' knowledge of religion, even their own traditions, is often sketchy, it may be prudent to hold back before freeing students to criticize. That said, there still needs to be some room for critical reflection, even in large introductory classes, lest the great divide between professors and students be widened. These are issues with which all RS programs need to contend. But higher education in general, and RS in particular, needs the presence of Socrates in the lecture hall. If not, then we execute Socrates all over again.

NOTES

1. This essay builds on many of the themes discussed in my earlier essay "The Socratic Method in Today's University," in Lee Trepanier, ed., *The Socratic Method Today: Student Centered and Transformative Teaching in Political Science* (New York: Routledge, 2018).
2. Barbara Walvoord, *Teaching and Learning in College Introductory Religion Courses* (Malden, MA: Blackwell, 2008).
3. Ibid., 2.
4. Ibid., 6.
5. See Christopher R. Cotter and David G. Robertson, eds., *After World Religions: Reconstructing World Religions* (New York: Routledge, 2016); Mark C. Taylor, ed., *Critical Terms for Religious Studies* (Chicago: University of Chicago Press, 1998).
6. See Jonathan Z. Smith, "Religion, Religions, Religious," in Taylor, *Critical Terms* (Chicago: University of Chicago Press, 1998), 269–84.
7. Walvoord, *Teaching and Learning*, 6.
8. Ibid., 5.
9. Ibid., 25–7.
10. Ibid., 35–7.
11. Cotter and Robertson, "Introduction: The World Religions Paradigm in Contemporary Religious Studies," in *After World Religions* (New York: Routledge, 2016), 11.
12. Walvoord, *Teaching and Learning,* 8–9.
13. Friesen, *The Textbook and the Lecture: Education in the Age of New Media* (Baltimore: John Hopkins University Press, 2017), 110–38.
14. For contemporary considerations of the Socratic Method in higher education, see Trepanier, ed., *Socratic Method Today.*

15. Plato, *Symposium*, 201d–7a. Also see Plato, *Apology*, 31d, 40a–c, where Socrates says that his *daemon* stops him from saying or doing things that are wrong.

16. For some examples of Socratic myths, see Plato, *Republic*, 614b–21d; *Gorgias*, 523a–7e; *Phaedrus*, 244a–57b; *Phaedo* 107c–15a.

17. Stephen Prothero, *God is Not One: The Eight Rival Religions that Run the World* (New York: HarperOne, 2010), 4.

18. Ibid., 3.

19. Levinovitz, *The Limits of Religious Tolerance* (Amherst, MA: Amherst College Press, 2016), 35.

20. Ibid., 26–7.

21. Cornell West speaks of the classroom as a space where "you thoroughly unsettle people." See Mark Bernstein, "In Defense of UnSafe Spaces," *Princeton Alumni Weekly,* December 7, 2016. Available at https://paw.princeton.edu/article/conversation-speaking-their-minds.

22. Levinovitz, *Limits,* 31.

23. Ibid., 32.

24. Cotter and Robertson, "Introduction," 6.

25. See Tomoko Masuzawa, *The Invention of World Religions, Or, How European Universalism Was Preserved in the Language of Pluralism* (Chicago: University of Chicago Press, 2005).

26. Plato, *Phaedrus*, 274b–7a.

27. Lemoine, "Is Socrates Culturally Imperialistic," in Trepanier, *Socratic Method Today*, 126.

28. Ibid., 131.

29. Ibid., 128–9, 132.

30. Laurie L. Patton and Vernon K. Robbins, with Gordon D. Newby, "Comparative Sacred Texts and Interactive Interpretation: Another Alternative to the 'World Religions' Class," *Teaching Theology and Religion* 12/1 (2009): 37–49.

WORKS CITED

Bernstein, Mark. 2016. "In Defense of UnSafe Spaces." *Princeton Alumni Weekly,* December 7. Available at https://paw.princeton.edu/article/conversation-speaking-their-minds.

Cooper, John M., ed. 1997. *Plato: Complete Works*. Indianapolis: Hackett Publishing Co.

Cotter, Christopher R. and David G. Robertson, eds. 2016. *After World Religions: Reconstructing World Religions*. New York: Routledge.

Friesen, Norm. 2017. *The Textbook and the Lecture: Education in the Age of New Media*. Baltimore: John Hopkins University Press.

Levinovitz, Alan J. 2016. *The Limits of Religious Tolerance*. Amherst, MA: Amherst College Press.

Masuzawa, Tomoko. 2005. *The Invention of World Religions, Or, How European Universalism Was Preserved in the Language of Pluralism*. Chicago: University of Chicago Press.

Patton, Laurie L., Vernon K. Robbins, with Gordon D. Newby. 2009. "Comparative Sacred Texts and Interactive Interpretation: Another Alternative to the 'World Religions' Class." *Teaching Theology and Religion* 12/1: 37–49.

Prothero, Stephen. 2010. *God is Not One: The Eight Rival Religions that Run the World.* New York: HarperOne.

Taylor, Mark C., ed. 1998. *Critical Terms for Religious Studies*. Chicago: University of Chicago Press.

Trepanier, Lee, ed. 2018. *The Socratic Method Today: Student Centered and Transformative Teaching in Political Science.* New York: Routledge.

Walvoord, Barbara. 2008. *Teaching and Learning in College Introductory Religion Courses*. Malden, MA: Blackwell.

Chapter 15

Curious Lectures and Engaged Students

Teaching in a Context of Bureaucracy and Consumerism

Warren Fincher

My strategic goals for the first lecture of the semester are clear: preview the course content and define the scope of the class, establish the operating norms of my classrooms, and set an engaging instructional tone. Examining goals of the first day of lecture lays bare the various social stakeholders that the instructor must satisfy in the course of fulfilling her duty: the discipline, the organization, and the student. But of these stakeholders, though, it is the student who sits in the classroom with the instructor, and the work to be accomplished within the class session involves *both* the instructor and the student.

I recall the moment, during the first year of my undergraduate education, when I realized that I should be as intellectually engaged during lectures as when reading for class, writing a paper, or studying for an exam (and the subsequent revelation that, if I indeed did treat lectures as a part of my studying rather than just a calendar obligation, then earning good marks on papers and exams became much easier). We cannot assume that students bring this attitude into the classroom with them, and it becomes at least partially incumbent on the instructor to find ways to foster the inquisitive mind of the student, alongside the other goals of instruction. In *The Child and the Curriculum*, the prominent educational reformer John Dewey acknowledges the interactional nature of classroom education, paying focused attention to balancing the educator's task of communicating information with the students' ability to assimilate that new information.[1] By extension, as much as education is co-created between the engaged student and the professing instructor, any alteration to the student's social context outside the classroom—in the twenty-first century, this might include the prevalence of social media,[2] the ease of streaming mass media,[3] the prevalence of online multiplayer gaming,[4] and a culture of consumerism[5]—may impact the demands on the instructor's pedagogy.

The course instructor is responsible for balancing the organizational, disciplinary, and interactional demands in creating an engaging course with specified content. Over the past decade or so, digital technologies have added a variety of new engagement techniques to the instructor's toolkits: videos as assigned "readings," collaborative work through file-sharing (including wiki constructions), and discussion boards to name a few. Still, though online education continues to enroll larger and larger shares of students, the classroom remains the primary vehicle through which university instruction occurs.[6] And while workshops, laboratories, service learning, simulations, and "flipped" instruction are increasingly salient methods of organizing instruction, the lecture persists as a key pedagogical tool in the classroom. Between the efficiency that the lecture provides, particularly with large enrollments, and the cultural momentum of teaching in the ways that instructors themselves were taught throughout their undergraduate years, the lecture persists.

Thus, the lecture is both a pedagogical tool and a social event, and this chapter will take the latter as its point of departure for examining how the lecture-as-tool might successfully respond to the educational setting of the twenty-first century, given administrative and accreditation demands on the one hand and the media and consumeristic demands over students' attention on the other. This chapter applies a role analysis approach, viewing the lecture as an interaction between two members of a role pair—the lecturer and the student—with the goal of successfully educating the student as defined by the standards of the discipline and accrediting organizations. In doing so, this discussion will speculate how methods of engagement might be borrowed from akin interactional forms—the informational business meeting and investigative mass media, for example—and set successfully within the university lecture. This chapter intends less to provide a set of instructional methods for lecturing and instead aspires to provide the reader with an analytic lens through which he or she might assess his or her own lectures and creatively reflect upon how his or her lecture might borrow from other interactional forms to heighten student engagement.

THE LECTURE AS SOCIAL INTERACTION

Defining the lecture shouldn't require great imagination to anyone who has advanced very far in their education; in its ideal-typological sense, the lecture as a social interaction involves an expert orating information to a collection of people who hold an interest in learning that information. The educational content is structured by disciplinary standards, which is why the lecturer must be an expert in the material at hand. This information may be objective in nature, focusing on tasks as cognitively simple as factual comprehension, or the lecture may work to foster more honed analytic and evaluative skills for the student. The role pair of lecturer and student also presumes that the latter, though she may refrain from speaking during the lecture, is nonetheless intellectually engaged. The well-formed question from the student can evidence a cognitive engagement, as might judicious note-taking. But aside from these behaviors, the role of the student is to conduct the intellectual work on their side of the lectern to internalize the information provided, understand its relevance to their educational goals, and engage analytic faculties that need further development.

This account serves as an ideal-type description of the lecture. Sociologists use ideal types as an abstraction of a social phenomenon that represents key features in an extreme or exaggerated manner; it is not intended to be a statistically commonplace depiction, but a philosophically pure representation that serves as an analytic tool for the researcher. While the above description isolates the specific roles involved in the social phenomenon of the lecture, it is stripped of important contextualizing features of the twenty-first-century educational milieu.

The content of the lecture is not simply a manifestation of the discipline voiced through the expert's oration, but rather it is structured by administrative concerns: the location of the lecture within a semester-long set of educational goals, within the curriculum of the major or minor or General Education program, and within the standards of both educational accrediting agencies, such as the Higher Learning Commission, and professional accrediting agencies that regulate business, medical, or educational degree programs. With these concerns at hand, the lecturer is not only performing knowledge work that meets disciplinary standards but is also performing work that meets bureaucratic standards. And commensurate with this, the lecturer must provide evidence of having met those standards; inherent today in the act of lecturing is the expectation that the lecturer ultimately engage in data collection and, thus, justify the lecture's appropriateness to, and effectiveness in, meeting certain administrative goals. This does not necessarily displace the disciplinary concerns over the lecture's content, but it does add to the lecturer's role, that of a bureaucrat.

Not only has the lecturer's role changed in response to the hand of university and accrediting bureaucracies reaching into the classroom, but the context of the university student also departs from the ideal-type profile of the learner. No longer is the university populated exclusively with residential students who can afford the cost of attending college. With 40 percent of undergraduates working more than thirty hours weekly while enrolled in college classes,[7] the attention and energy required of students to learn from college classes is increasingly have to be pried away from the demands of the student's job(s). It should perhaps be no surprise, given the extent to which students concurrently pursue a college degree and work jobs, that many students may become disinterested in course content that they do not perceive as directly relevant to their future careers. How often do those of us who teach in general education programs hear complaints from students about the perceived irrelevance of the course material to them? Another departure of the twenty-first-century student from that of the ideal-typical described above involves the consumerist attitudes taken on by students.[8] The university is an experience for which the student pays handily, so why shouldn't he or she feel entitled to the benefits afforded the typical consumer? This issue is reflected in the classroom as some students feel that basic participation in the course should warrant a high grade[9] and outside of the classroom through websites like ratemyprofessors.com, which host user-reviews of faculty and their classes.[10]

How does the lecturer respond to the demands of her paired role, that of the pragmatic, consumeristic student, amid an administrative context that imposes bureaucratic standards upon the lecture's content? Is the lecturer a scholar, or is she a bureaucrat, or is she a producer set on selling her wares to skeptical consumers? We might borrow from Eliot Freidson the conceptual framing of the "professional" to better understand the countervailing influences over the lecturer-student role pair.[11] For Freidson, the professional stands at the intersection of expertise, bureaucracy, and markets. The professional gains some degree of market protection (or "enclosure") from the state because the kinds of knowledge needed to perform the task at hand is held only by those with specific training; not everyone has the knowledge to be a surgeon, for example, and the state has an interest in making sure that only those properly trained provide surgical services. The certification of that knowledge, though, becomes housed in a bureaucratic governance structure that is often partially dictated by the state (legal minimums to practice surgery, for example) and otherwise organized by the community of experts themselves (such as the American Medical Association). University lecturers fall into an akin situation: most states require some minimum level of graduate work in order to teach a university-level course, thus creating a weak enclosure, while educational accrediting bodies ensure through periodic review that the lecturer (via assessment of the university as a whole) maintains appropriate outcomes.

If the lecturer may be framed as a professional, then the student correspondingly becomes the "client" within the role pair. Clients are people who seek assistance from those certified to have specialized knowledge or skill. To the extent that customers have the latitude to make specific demands in the marketplace, clients do not have this same authority—by definition they do not have the authority held by the professional. Clients cannot demand certain medical procedures or pharmaceuticals from a doctor, illegal financial manipulations from an accountant, or code violations from an architect.

Similarly, the student does not have the right to demand an unwarranted grade from an instructor. Rather, the student has contracted with the instructor for a service, to be guided through educational material as part of a course of study. Much like the patient must follow through with a medical regime or the tax client must maintain and forward financial information to the accountant, the student also must play an active role in their own education: reading or viewing assigned materials, maintaining intellectual engagement during class sessions, studying for exams. The instructor's job, as an expert in the field, is to provide the structure for that course of study and to serve as a resource for the student when he or she needs assistance, but as with the professional-client role pair, the student may not remain a passive consumer of a service and expect to receive the fullest extent of that service's benefit. Education requires both instructor and student to work collaboratively for the service provided to meet a successful end.

The lecture serves as one of many pedagogical tools available to the instructor as she provides educational guidance to the student. It is not a perfect tool. Yes, it is an efficient way to communicate volumes of information within a limited time to a sizable number of students; bureaucratically and fiscally, the lecture is a very satisfying pedagogical tool. But the lecture, often a unidirectional oration given by the instructor toward the students, leaves the students tacitly in a passive position, often providing little assistance to students in maintaining an active engagement. Looking over my students' notes from time to time, I see evidence that they do not understand this role. Often, I will see a replication of my presentation slides but no mention of the material I presented only through oration. And rarer still are questions in the margins or documentation of a flight of thought inspired by something presented in lecture. As a pedagogical tool, shouldn't the lecture necessarily attend to supporting the student's role as an active learner? This goal comes more naturally in laboratories, workshops, studios, simulations, and seminars than in the lecture, but I believe that instructors can, if they carefully plan their lectures, ease the burden on students to remain actively engaged in their own educations.

LET'S NOT REINVENT THE WHEEL—LET'S STEAL ONE INSTEAD

Many presentational tools that have been honed to elicit viewer engagement can be found in the more bureaucratically organized or consumer-oriented milieux. An examination of these other unidirectional presentational forms might reveal techniques or strategies that can be employed in the classroom lecture in such a way as to support student engagement without challenging the efficiencies that the lecture affords.

Business Meetings. Not all business meetings are unidirectional. In fact, many of the tips given for effective workplace meetings specifically advise giving a role or task to everyone in attendance and to allow time for their participation.[12] But heavily interactional pedagogies can reduce many of the efficiencies of the lecture. Still, there are insights from the successful business meeting that can be incorporated into the lecture. Effective business meetings foster an environment in which people can ask for clarifications and suggest unconventional solutions to problems without reprisal. Guidance that the meeting facilitator fosters a safe environment for risk-taking also has applicability for the lectures, as students who fear asking for clarification may become hopelessly lost as the lecture proceeds or otherwise disengage themselves from the lecture altogether. Asking for possible interpretations of situations or solutions to problems can also mimic the kind of problem solving found in business meetings.

Good business meetings also communicate a clear purpose and set of goals for the meeting. Identifying the educational goals for a lecture, setting the lecture into the broader arc of the semester, and reviewing the lecture goals at the end of the class session are all ways to communicate to students their task within the lecture. Good business meetings should close with a clear sense of what the employee's next steps will be in completing a project; translating this to the lecture, the instructor might pre-teach assigned readings for the subsequent lecture, identifying the important elements to comprehend and clarifying how the reading relates to the educational arc of the course. These strategies stand to foster greater engagement with the lecture as students better understand their tasks within the co-production of education.

Games

Much can be said about the trend toward the gamification within education,[13] but my comments here focus on the gamification of the lecture, and specifically game scenarios that maintain the lecturer as the primary voice within the classroom while fostering attentiveness and engagement on the part of the students. One commonplace method involves providing the student with lecture notes that have omitted specific terminology or concepts, expecting the student to fill these in as she spots the missing information within the lecture; this is not unlike the game show, in which the viewer may attempt to solve a puzzle or answer a question alongside the contestants on the show. This method becomes more than just an Easter egg hunt when the missing information requires more than simply recognizing a specific term having been used in lecture but asks the student to employ higher analytic skills in order to complete the task at hand—rather than have a student listen for the lecturer to use the term "quiet racism," for example, she is instead asked to determine which of the examples provided is illustrative of quiet racism. This can be made a communal exercise with the use of audience response systems like iClicker, through which students can record their responses to questions that quiz on facts, seek opinions, or survey for shared experiences. The lecturer has the opportunity to address the aggregated response of the class without losing the efficiency of the unidirectional lecture; the student has the opportunity to safely voice an answer to a question, receive immediate feedback, and compare her thoughts or experiences to those of her classmates.

Educational Entertainment

Though often associated with fictional narratives and reality television, mass media can serve an educational purpose. Certainly instructional videos on platforms such as YouTube, expositions found in TED Talks, and polished documentary films provide the university students with entertaining educational experiences. What presentational tips might we steal from edutainment? The strategy of showing, rather than telling, seems relevant. The often text-heavy nature of lecture slides does not actively promote intellectual engagement, serving more as a visual duplicate of the orated speech or even a stand-in for the lecturer herself. Slides with minimal text but engaging visual images provide more of an evocative backdrop for the lecturer, allowing the ears and eyes to connect the lecturer's speech to a visual supplement, perhaps an illustrative diagram that serves as a map of the lecture or an image that helps transport the student's imagination outside of the classroom to another time or place.

Edutainment media, particularly successful documentary films, also often address relevant values and involve the audience's emotions. The past decade of Academy Award–winning documentary features consistently address issues of societal values and all attempt to elicit emotional interpretations alongside intellectual ones. If not passion, at least interest is piqued by engaging value-oriented and affective facets of any given subject. And, through visual depictions and appeals to emotion and values, educational entertainment has the capacity to connect the concrete, specific topic at hand to abstract principles and macrosocial trends.

Investigative News Media

Personally, I am not a fan of the nightly newscast, neither national nor local. The broadcasts seem scattershot in their coverage, providing too little time to deeply explore any one topic and offering a collection of news items that seem disconnected to each other, and I suspect that the nightly news does not provide a good template for enhancing student engagement in lectures.

However, younger Americans seem to be taken with hour-long news programs that feature punditry—in 2018 "The Rachel Maddow Show" had earned top viewership levels, notable in attracting younger viewers—and they also consume podcasts featuring in-depth investigations, either providing single-episode coverage as with "The Daily" or multiple-episode explorations as with "Serial." Some theorize that it is the framing of absurdity and use of irony that typifies Rachel Maddow's style,[14] but I suspect part of the appeal for younger viewers of Maddow and listeners of investigative podcasts is the way in which the narrative is layered such as to heighten the audience's curiosity about where the narrative is heading. Maddow often begins her show with some point of history that is seemingly insignificant to the day's news, but she uses this discussion to address the complex backstory of a current issue in the news. This connect-the-dots style that elicits curiosity is supplemented by a repetitiveness that makes sure the audience doesn't lose a sense of the connections between those dots. Podcasts like "the Daily" take a similar approach, though the narratives often quickly step away from directly addressing the issue at hand and instead explore the complexity that stands behind the news story, creating questions and answering them with the next action taken by those involved, the next quotation from a source, the next interview with an expert. The commonality seems to be building questions into the narrative, answering the questions with evidence, and continually outlining the narrative so that the audience doesn't lose the narrative.

FINAL REFLECTIONS

Space constraints limit the number of presentational forms discussed here, but hopefully the examination of these four have provided a starting point for exploring the potential that other lecture-like communicative forms hold in making your lectures more engaging—and ultimately supportive of the active-learner role that lecture requires of the student. The work of the university lecturer must respond to many organizational and societal pressures, but the interactive relationship between the lecturer and the student is essential, as is recognizing that the student cannot be a passive actor in this work. The successful lecture must support student engagement and student curiosity.

Perhaps you've identified some of your own lecturing strategies in the above conversation and maybe you've picked up a few items to employ in your next lecture, but more importantly, I hope that this discussion has set out a line of flight for creatively and analytically thinking about how the lecture can borrow from akin interactional forms and media styles to help facilitate the student's role in remaining intellectually engaged in a pedagogical style that can too easily become passive on the part of the student.

NOTES

1. Dewey, *The Child and the Curriculum* (Chicago: University of Chicago Press, 1902).
2. Junco, "Too Much Face and Not Enough Books: The Relationship between Multiple Indices of Facebook Use and Academic Performance," *Computers in Human Behavior* 28 (2012): 187–98; and Karpinski et al., "An Exploration of Social Networking Site Use, Multitasking, and Academic Performance Among United States and European University Students," *Human Behavior* 29 (2013): 1182–92; and Tess, "The Role of Social Media in Higher Education Classes (Real and Virtual) – A Literature Review," *Computers in Human Behavior* 25/5 (2013): A60–A68.
3. Calderwood et al., "What Else Do College Students 'Do' While Studying? An Investigation of Multitasking," *Computers & Education* 75 (2014): 19–29; and Portanova, "The Rhetoric of Distraction: Media Use and the Student Writing Process," in *Social Writing/Social Media: Publics, Presentations, and Pedagogies*, Douglas Walls and Stephanie Vie eds. (Fort Collins, Colorado: WAC Clearinghouse and University Press of Colorado, 2017), 249–64.
4. Duggan, "Gaming and Gamers," (Pew Research Center, 2015); and Lenhart et al.,"Teens, Video Games, and Civics," (Pew Internet and American Life Project, 2008).
5. Titus, "Student Ratings in a Consumerist Academy: Leveraging Pedagogical Control and Authority," *Sociological Perspectives* 51/2 (2008): 397–422.
6. Seaman et al., "Grade Increase: Tracking Distance Education in the United States," (Babson Survey Research Group, 2018).
7. Carnevale et al., "Learning While Earning: The New Normal," (Georgetown University Center on Education and the Workforce, 2015): 11.
8. Schaffer et al., "The 'You Owe Me!' Mentality: A Student Entitlement Perception Paradox," *Journal of Learning in Higher Education* 9/1 (2013): 79–91.
9. Greenberger et al., "Self-Entitled College Students: Contributions of Personality, Parenting, and Motivational Factors," *Journal of Youth and Adolescence* 37/10 (2008): 1193–204.
10. Davison and Price, "How Do We Rate? An Evaluation of Online Student Evaluations," *Assessment and Evaluation in Higher Education* 34/1 (2009): 51–65; and Titus, "Student Ratings in a Consumerist Academy"; and Yoon "Affordances and Negotiations of the Digital Reputation Society: A Case Study of RateMyProfessors.com," *Continuum Journal of Media & Cultural Studies* 29/1 (2014): 109–20.
11. Freidson, *Professionalism: the Third Logic* (Chicago: University of Chicago Press, 2001).
12. Bryant, "How to Run a More Effective Meeting," *New York Times*, April 24, 2017. Available at https://www.nytimes.com/guides/business/how-to-run-an-effective

-meeting; and Harvard Business Review, *HBR Guide to Making Every Meeting Matter* (Brighton: Harvard Business School Press, 2016).

13. Chou, *Actionable Gamification: Beyond Points, Badges and Leaderboards* (Milpitas, California: Octalysis Media, 2015); Kapp, *The Gamification of Learning and Instruction: Game-Based Methods and Strategies for Training and Education* (San Francisco: Pfeiffer, 2012).

14. Quart, "The Sarcastic Times: For Rachel Maddow and the Other Ironic Anchors, Absurdity is Serious Stuff," *Columbia Journalism Review*, March/April 2009. Available at https://archives.cjr.org/essay/the_sarcastic_times.php.

WORKS CITED

Bryant, Adam. 2017. "How to Run a More Effective Meeting." *New York Times*, April 24. Available at: https://www.nytimes.com/guides/business/how-to-run-an-effective-meeting.

Calderwood, Charles, Phillip L. Ackerman, and Erin Marie Conklin. 2014. "What Else Do College Students 'Do' While Studying? An Investigation of Multitasking." *Computers & Education* 75: 19–29.

Carnevale, Anthony P., Nicole Smith, Michelle Melton, and Eric W. Price. 2015. "Learning While Earning: The New Normal." Georgetown University Center on Education and the Workforce. Available at: https://1gyhoq479ufd3yna29x7ubjn-wpengine.netdna-ssl.com/wp-content/uploads/Working-Learners-Report.pdf.

Chou, Yu-kai. 2015. *Actionable Gamification: Beyond Points, Badges and Leaderboards*. Milpitas, California: Octalysis Media.

Davison, Elizabeth and Jammie Price. 2009. "How Do We Rate? An Evaluation of Online Student Evaluations." *Assessment and Evaluation in Higher Education* 34/1: 51–65.

Dewey, John. 1902. *The Child and the Curriculum*. Chicago: University of Chicago Press.

Duggan, Maeve. 2015. "Gaming and Gamers." Pew Research Center. Available at http://www.pewinternet.org/2015/12/15/gaming-and-gamers/.

Freidson, Eliot. 2001. *Professionalism: The Third Logic*. Chicago: University of Chicago Press.

Greenberger, Ellen, Jared Lessard, Chuansheng Chen, and Susan P. Farruggia. 2008. "Self-Entitled College Students: Contributions of Personality, Parenting, and Motivational Factors." *Journal of Youth and Adolescence* 37/10: 1193–204.

Harvard Business Review. 2016. *HBR Guide to Making Every Meeting Matter*. Brighton: Harvard Business School Press.

Junco, R. 2012. "Too Much Face and Not Enough Books: The Relationship between Multiple Indices of Facebook Use and Academic Performance." *Computers in Human Behavior* 28: 187–98.

Kapp, Karl M. 2012. *The Gamification of Learning and Instruction: Game-Based Methods and Strategies for Training and Education*. San Francisco: Pfeiffer.

Karpinski, A.C., P.A. Kirschner, I. Ozer, J.A. Mellott, and P. Ochwo. 2013. "An Exploration of Social Networking Site Use, Multitasking, and Academic Performance Among United States and European University Students." *Human Behavior* 29: 1182–92.

Lenhart, Amanda, Joseph Kahne, Ellen Middaugh, Alexandra Rankin Macgill, Chris Evans, and Jessica Vitak. 2008. "Teens, Video Games, and Civics." Pew Internet and American Life Project. Available at: http://www.pewinternet.org/wp-content/uploads/sites/9/media/Files/Reports/2008/PIP_Teens_Games_and_Civics_Report_FINAL.pdf.pdf.

Portanova, Patricia. 2017. "The Rhetoric of Distraction: Media Use and the Student Writing Process." In *Social Writing/Social Media: Publics, Presentations, and Pedagogies*, Douglas Walls and Stephanie Vie, eds. Fort Collins, Colorado: WAC Clearinghouse and University Press of Colorado. 249–64.

Quart, Alissa. 2009. "The Sarcastic Times: For Rachel Maddow and the Other Ironic Anchors, Absurdity is Serious Stuff." *Columbia Journalism Review*. Available at https://archives.cjr.org/essay/the_sarcastic_times.php.

Schaefer, Thomas, Marguerite Barta, William Whitley, and Margie Stogsdill. 2013. "The 'You Owe Me!' Mentality: A Student Entitlement Perception Paradox." *Journal of Learning in Higher Education* 9/1: 79–91.

Seaman, Julia E., I. Elaine Allen, and Jeff Seaman. 2018. "Grade Increase: Tracking Distance Education in the United States." Babson Survey Research Group. Available at https://onlinelearningsurvey.com/reports/gradeincrease.pdf.

Tess, Paul A. 2013. "The Role of Social Media in Higher Education Classes (Real and Virtual) – A Literature Review." *Computers in Human Behavior* 25/5: A60–A68.

Titus, Jordan J. 2008. "Student Ratings in a Consumerist Academy: Leveraging Pedagogical Control and Authority." *Sociological Perspectives* 51/2: 397–422.

Yoon, Kyong. 2014. "Affordances and Negotiations of the Digital Reputation Society: A Case Study of RateMyProfessors.com." *Continuum Journal of Media & Cultural Studies* 29/1: 109–20.

Index

About the Authors

Karen P. Burke is an associate professor of media studies at Southern Connecticut University. She has published numerous articles and chapters in media studies.

Paul Corey is a professor of religious studies at Humber College in Canada. He is author of *Messiahs and Machiavellians: Depicting Evil in Modern Theatre* (Notre Dame, 2008).

John Craig is a professor of political science at Leeds Beckett University in Great Britain. He is a National Teaching Fellow, co-convener of the UK Political Studies Association Teaching and Learning Network, and a member of the editorial board for the *Journal of Political Science Education.*

Monika Dix is an associate professor of Japanese language, literature, and culture at Saginaw Valley State University in Michigan. She has published numerous articles and chapters on Japan and currently working on a manuscript on Buddhism in Japanese literature.

Warren Fincher is an associate professor of Sociology at Saginaw Valley State University in Michigan.

Jamie Fiste is an associate professor of music at Central Michigan University. He is an active cellist, performing internationally, as well as has written essays on music pedagogy.

Eric Michael French is an assistant professor in political science at Oklahoma State University.

Emily Kelley is an associate professor of art history at Saginaw Valley State University in Michigan. Her most recent publication is *Saints as Intercessors between the Wealthy and the Divine: Art and Hagiography among the Medieval Merchant Classes* (Routledge, 2019).

Paul Krause is a doctoral student in religious studies at the University of Buckingham in Great Britain. He received his MA in theology at Yale University and has published numerous essays on religion and culture.

Hyrum Lewis is a professor of history at Brigham Young University at Idaho. He is author of several books, the latest being *American Foundations: Politics, Economics, Culture* (BYU, 2017).

Mike Mosher is a professor of media studies at Saginaw Valley State University in Michigan. He is author of several books, the latest being, with David Smith, *George Orwell Illustrated* (Haymarket Books, 2018).

Taryn Okuma is a professor of English at Catholic University of America in Washington D.C. She has published numerous articles and chapters on Henry Green, Evelyn Waugh, Kazuo Ishiguro, W. G. Sebald, T. S. Eliot, Pearl Buck, and college composition.

Elizabeth Rich is a professor of English and director of the Honors Program at Saginaw Valley State University in Michigan. She has published numerous articles and chapters in literary criticism and theory.

Caroline R. Sherman is a professor of history at Catholic University of America in Washington D.C. She is author of *Uses of the Dead: The Early Modern Development of Cy-Près Doctrine* (Catholic University of America, 2018).

Lee Trepanier is a professor of political science at Saginaw Valley State University and editor of Lexington Books series Politics, Literature, and Film and the academic website, *VoegelinView*. He is the author and editor of over twenty books, the latest being *The Socratic Method Today: Student-Centered Learning and Transformative Teaching in Political Science* (Routledge, 2018).

Brendon Westler is a postdoctoral research fellow with the Center for the Study of Liberal Democracy at the University of Wisconsin-Madison.